Climate Governance
at the Crossroads

Climate Governance at the Crossroads

EXPERIMENTING WITH A GLOBAL RESPONSE AFTER KYOTO

MATTHEW J. HOFFMANN

OXFORD
UNIVERSITY PRESS

OXFORD
UNIVERSITY PRESS

Oxford University Press, Inc., publishes works that further
Oxford University's objective of excellence
in research, scholarship, and education.

Oxford New York
Auckland Cape Town Dar es Salaam Hong Kong Karachi
Kuala Lumpur Madrid Melbourne Mexico City Nairobi
New Delhi Shanghai Taipei Toronto

With offices in
Argentina Austria Brazil Chile Czech Republic France Greece
Guatemala Hungary Italy Japan Poland Portugal Singapore
South Korea Switzerland Thailand Turkey Ukraine Vietnam

Copyright © 2011 by Oxford University Press, Inc.

Published by Oxford University Press, Inc.
198 Madison Avenue, New York, NY 10016

www.oup.com

Oxford is a registered trademark of Oxford University Press

Library of Congress Cataloging-in-Publication Data
Hoffmann, Matthew J.
Climate governance at the crossroads: experimenting with a global response
after Kyoto / Matthew J. Hoffmann.
    p.   cm.
Includes bibliographical references and index.
ISBN 978-0-19-539008-7
1. Environmental policy—Government policy.
2. Environmental policy—International cooperation.
3. Climatic changes—Government policy.
4. Climatic changes—International cooperation. I. Title.
GE170.H64 2011 363.738'74526—dc22 2010023217

9 8 7 6 5 4 3 2 1

Printed in the United States of America
on acid-free paper

Contents

Tables and Figures

Tables

Figures

Preface

The research for this book emerged from a confluence of disillusionment and serendipity. I spent most of my early career approaching the study of global environmental politics by examining the multilateral treaty negotiations at the center of the world's response to environmental challenges like ozone depletion and climate change. But beginning around 2003–2004, I became increasingly disillusioned with the multilateral process both personally and academically. Personally, I was becoming more and more frustrated with and concerned about the lack of progress in the addressing climate change—becoming a new father had brought the problem of climate change home to me in a different light than I had perceived it before and, frankly, I was (and remain) scared of the potential consequences of climate change. Academically, I was frustrated as well. It seemed that academics had adequately diagnosed the reasons collective action was not forthcoming on climate change, but that we of the academy (political scientists, economists, environmental studies scholars), collectively, were having a difficult time breaking through the hold that multilateral treaty-making had on our own and on policy-makers' imaginations. Fortunately, this was not universally the case. At around the same time, some pioneering studies were beginning to examine the global response to climate change from a broader perspective, and I drew inspiration from much of this work on cities, NGOs, and corporations, as will be obvious in the pages that follow.

Yet it took serendipity to turn my general discontent in something more productive in 2007–2008. One day in the spring of 2007, my morning paper had a story about carbon rationing action groups—small, loosely affiliated groups of people in the United Kingdom (at the time—now the phenomenon has spread to the United States, Canada, and China) that were negotiating and imposing Kyoto Protocol–like restrictions on themselves at an individual level. Here you had extremely micro initiatives drawing on motifs and modes of governance that nation-states employ. This struck a chord with me and was a particularly crystallizing moment. I distinctly remember saying to myself, "They're experimenting." They are trying something out that works for them, but with a hope that there will be some larger ramifications. This drive to innovate, to try a different way of responding to climate change (not necessarily the details of what was happening in these particular groups, though it is a fascinating case) was something I wanted to understand and know more about and I wanted to see if there was any potential in experimentation to alter the course of the global response to climate change.

So I began a process of gathering information on as many unconventional initiatives as I could. Aided by an extremely talented and dedicated research assistant—Gabe Eidelman—I devised and revised criteria for identifying the 58 climate governance experiments analyzed in this book. I worked on putting together a database and research project to further understand what I came to consider an experimental system of governance—how and why individual experiments have emerged and how they are interacting and influencing the global response to climate change. In addition to gathering data from Web sites and documents to fill the database and lay the foundation for examining the collection of experiments, the research took me to the Conference of the Parties meetings for the UN climate change negotiations in 2007 (in Bali) and 2009 (in Copenhagen). I also attended the Carbon Markets Insights America conference hosted by the Point Carbon in November 2009. Finally I undertook over 40 interviews with individuals who were actively involved in climate governance experiments.

In the course of research and writing, it became clear to me that what I wanted to accomplish with this project was both an academic study of climate governance experiments and a more practically oriented exploration of the global response to climate change that is emerging through climate governance experiments. My hope for this book is that it will contribute to the burgeoning literature on global environmental governance but also be a bit helpful for people working on climate change on the ground like those who were gracious enough to speak with me about their work addressing climate change. The initiatives I explore in this book are all relatively new and are experimental in the sense that we cannot yet be sure how they will turn out. By publicizing (in however small a way an academic book can do) this activity taking place outside the traditional multilateral channels and by illuminating how these initiatives are forming the nascent basis for a coherent experimental system of governance, I hope to aid their work.

Acknowledgments

As with any project of this size, it was not accomplished alone. Gabe Eidelman's assistance was crucial to the success of this book substantively and even stylistically (he is a whiz with graphics). Dave Gordon, Adam Senft, and Roberta Bell also provided valuable research assistance for various portions of the manuscript. I have benefitted enormously from collaboration with Michele Betsill on a project tracking emissions trading mechanisms, and she has also suffered through my excitement about this project from its inception and provided wonderful feedback. Along with Michele, Steven Bernstein, Matthew Paterson, and I are also engaged in a long-term research project on the governance and legitimacy of carbon markets. Working with these outstanding scholars has been a privilege, and I have learned a great deal in the process that has improved my work on this project. I owe debts of gratitude to a number of people for providing feedback on ideas and chapters along the way. Greg Hoffmann and John Heugel were gracious enough to read the manuscript from a nonacademic perspective and help me to keep the academic jargon to a minimum. All the feedback and encouragement I received from Steven Bernstein, Michele Betsill, Lilach Gilady, Harriet Bulkeley, Dave Gordon, Gabe Eidelman, Teresa Kramarz, David Levy, Elinor Ostrom, Mat Paterson, Phil Triadafilopoulos, Stacy VanDeveer, and the members of the Leverhulme research network on Transnational Climate Governance has been instrumental in improving my ideas and writing—and all these persons, of course, are blameless for any faults of logic or fact that remain. My editor at Oxford University Press, Angela Chnapko, was wonderfully encouraging and provided both feedback and support through the writing and publication process.

Working on this project and thinking about climate governance experimentation reminded me again of my intellectual debts to my longtime mentors Marty Finnemore and James Rosenau. Their ideas, training, and passion will always be crucial resources I draw on, and their insights are to be found throughout this project.

I could not have written this book without the support of my friend, partner, and wife, Lena Mortensen. She is my biggest cheerleader and my staunchest critic, and she makes everything I do better. It is clichéd, but nonetheless true, that I wrote this book for my son, Anders. This is perhaps not the book he envisioned when he asked me to write a book for him as a four-year-old in 2008, but I hope it is one he someday finds worthy. Anders is the reason I wake up at night worried about climate change, and writing this book is a small contribution I can make to ensuring that his future can be a bright one.

Climate Governance
at the Crossroads

1

Into the Void

Aren't we a little too self-righteous to pretend that all strategy is here in the toolbox of Kyoto, where there are only numerical target, timeline, some flexible mechanisms and detailed punishment plan? Shouldn't we be a little more humble to the awesome might of nature and human action and start exploring many more tools and strategies on top of the Kyoto's tool box?
—Japanese Submission for the 2006 Conference of the Parties to the United Nations Framework Convention on Climate Change in Nairobi

A Tale of Two Copenhagens

The world came to Copenhagen in December 2009, physically and virtually.[1] The focal point was the annual installment of the ongoing United Nations negotiations aimed at achieving an international agreement to combat climate change. These yearly "Conferences of the Parties" (COPs) galvanize interest in climate change, serving as the centerpiece of the international community's response to climate change since the original UN negotiations in the early 1990s. But UN negotiations were not the only show in town. There were two Copenhagens.

One was familiar and obvious, located at the Bella Conference Center that housed the official negotiations. Here activists paraded and shouted, festooned in eye-catching costumes, and berated the "Fossils of the Day" for obstructing progress on a global accord. They sought to catch negotiators in the corridors and press their positions, at least before the vast majority of these observers representing civil society from around the world were unceremoniously denied access in the closing days of the conference. The harried and

3

exhausted negotiators diligently worked long hours over multiple and conflict-ing versions of negotiating texts that revealed myriad fault lines dividing the nations of the world. Dozens of organizations maintained informational booths seeking to publicize their version of climate-friendly activities. Stu-dents and academics huddled in small groups trying to make sense of an enor-mous and unwieldy negotiating process, the most important aspects of which take place behind closed doors. Cameras and microphones were ubiquitous as the media sought both the main storyline of the negotiations and the smaller personal interest stories that would connect the global summit to peoples' lives back home. A logistical and security nightmare unfolded as heads of state swooped in during the final days of the conference. They sought a break-through compromise, but left with what many consider to be a disappointing Copenhagen Accord that does little to ensure that significant actions will be taken to address climate change.[2] Outside the Center, demonstrators chas-tised the negotiators and urged them to take action on climate change and equity concerns. The eyes of the world focused on the events at the Bella Center. Many now despair at what they witnessed.

The other Copenhagen manifested at multiple sites throughout the city.[3] One scant subway stop south of the Bella Center stands the Crowne Plaza Hotel, where the International Emissions Trading Association held a series of presentations and discussions—"side events" to the focal negotiations up the road.[4] The pace and feel of the Crowne Plaza was calm and relaxed in contrast to the frenetic atmosphere of the Bella Center. In two medium-sized conference rooms, representatives from various organizations—banks, corporations, car-bon traders, NGOs, think tanks, even nation-states—laid out how existing car-bon markets function and the plans for developing and scaling them up in an attempt to address climate change and make profit. In downtown Copenhagen, mayors, governors, and corporate leaders met with much fanfare at two con-current events. The Climate Summit for Mayors, sponsored by the C40 Cities Climate Leadership Group (hereafter C40) explored municipal responses to cli-mate change, and the Climate Leaders Summit highlighted the activities of members of The Climate Group. The Climate Leaders Summit produced a far-reaching agreement among subnational leaders (in less than an hour) that promises far more stringent action than is included in the Copenhagen Accord.[5] The eyes of the world were not tightly focused on this other Copenhagen: the Crowne Plaza, the leaders' summits, or similar venues across Copenhagen that showcased alternative approaches to addressing climate change. They should have been.

The two Copenhagens represent two very different ways of responding to climate change—devising the mechanisms, technologies, and institutions through which the world attempts to mitigate the sources and adapt to the ef-fects of human-induced climatic changes and global warming. The Bella Center

Copenhagen exemplified the traditional and familiar megamultilateral approach whereby all (or most) of the world's nation-states convene to negotiate a legally binding treaty that shapes the domestic actions of individual nation-states. This is top-down governance designed to smoothly transition from international co-operation to domestic implementation, an approach that makes a good deal of sense, as we are told time and again that climate change is a problem that cannot be solved by any single nation-state. The 1997 Kyoto Protocol is the most recent treaty to emerge from this means of responding to climate change, and the activities at the Bella Center Copenhagen sought to produce a replacement ahead of its expiration in 2012.

The other Copenhagen was something less familiar, messier, more diffuse and dynamic—in a word, *experimental*. The other Copenhagen revealed how cities, counties, provinces, regions, civil society, and corporations are responding to climate change independently from, or only loosely connected to, the "official" UN-sponsored negotiations and treaties. This is bottom-up governance, whereby myriad actors inspired by frustration with the multilateral process (deeming it either too slow or too fast), a sense of urgency about climate change, and even profit and power refuse to leave the response to climate change solely to the multilateral treaty negotiations—they are taking climate change into their own hands. Less focused on a singular outcome (a global treaty), these initiatives push the global response to climate change in a number of directions—energy efficiency, carbon markets, local adaptation, transformation of the built environment and transportation systems, among others.

The Bella Center Copenhagen is comfortable, if disappointing. This mode of climate governance has been the focus of the global response to climate change (and other transnational problems) for decades.[6] Yet the focus on the Bella Center in December 2009 obscured the importance of the other Copenhagen, just as the focus on multilateral treaty-making obscures the importance of bottom-up processes that have begun to percolate in the last decade. This book seeks to correct that imbalance. I demonstrate how the center of gravity in the global response to climate change is shifting from the multilateral treaty-making process to the diverse activities in the other Copenhagen. In the pages that follow I examine the development and functioning of this experimental world of climate governance and its relationship with the traditional multilateral response to climate change. While experiments in responding to climate change are new and unproven, they may represent the best hope for effectively responding to the climate crisis.

Such hope would be a welcome change after two decades of difficult multilateral negotiations.[7] By most accounts, the controversial Copenhagen Accord that emerged in the final hours of the negotiations is a failure. To be fair, there were positive results. Major developing countries have agreed to some monitoring and verification of their climate abatement activities, and countries in the global North pledged serious funding for adaptation and mitigation efforts in developing countries. Yet there are no binding commitments to reduce green-house gas emissions in the short to medium term—countries will be able to set and report their own domestically derived commitments.[8] In addition, nation-states and the climate change secretariat that oversees the negotiations are scrambling to reconcile the Copenhagen Accord with the other long-established negotiations based on the 1992 UN Framework Convention on Climate Change (UNFCCC) and the 1997 Kyoto Protocol. Funding and verification are steps forward, but the Copenhagen Accord adds to the confusion surrounding the multilateral response to climate change. Further, it contains little that moves us beyond what the international community agreed to do in 1992—develop national action plans, report emissions, and vaguely commit to stabilize greenhouse gas concentrations below dangerous levels. Two decades with very little tangible progress in international treaty-making.

The disappointment at the Bella Center should have been expected.[9] World leaders were awfully frank in their doubts in the lead-up to Copenhagen about what could be accomplished. Activists, academics, negotiators, and the public at large had set their sights on a legally binding replacement for the Kyoto Protocol ever since the 2007 UNFCCC COP negotiations in Bali produced a "Roadmap" designed to achieve a new treaty within two years. Yet the gulfs separating key countries were plain, and world leaders strove to set lowered expectations in the fall of 2009. An effective, legally binding international treaty was a highly improbable outcome at Copenhagen, regardless of the urgency of the problem or the focus of the world on "Hopenhagen" in December 2009.

Further, the so-called failure of the Copenhagen conference was not merely a failure of political will, as it is often cast in the media. Blame has been leveled variously at the United States and China for torpedoing the negotiations, and even Canada came under fire for being obstructionist, "winning" the Fossil of the Year designation,[10] but it is more useful to ask if the structure of these negotiations is the problem. The megamultilateral mode of responding to global environmental problems has long been stymied in climate change. The 1992 UNFCCC and 1997 Kyoto Protocol have been roundly criticized for failing to produce an effective response to the climate crisis, and the ongoing UN-sponsored negotiations have been variously characterized as stagnant, ossified, stalemated, and at an impasse.[11] The results of the Copenhagen conference did nothing to turn the tide of poor negotiating results, and subsequent multilateral activity in 2010 was similarly disappointing. It is now entirely unclear whether the megamultilateral process will ever be able to deliver the deep cuts in greenhouse gas emissions that the international scientific community warns are required to avert the most serious impacts of climate change.[12]

The silver lining of the failure at Copenhagen is that our attention and energy may be fruitfully redirected. Failure at the Bella Center makes the activities in the other Copenhagen all the more important for understanding and even enhancing the global response to climate change. Bottom-up initiatives representative of the other Copenhagen—experiments in responding to climate change—abound:

- *Carbon rationing action groups (CRAGs)* are transnationally linked local community groups in the United Kingdom, United States, and Canada that negotiate and impose Kyoto-like carbon emission reductions on themselves. In a remarkable example of self-organization, carbon rationing action groups have sprung up in the United Kingdom (32), the United States (4), Canada (2), and China (1).[13] These groups are considered by their members to be a reaction to the stalemate at the multilateral level. A member of one CRAG in Glasgow noted that he is a member "for many reasons but perhaps most importantly because it allows me to do at a local scale what I think our governments should be doing at a global scale."[14]
- *Corporate climate responsibility* has grown in importance. In the 1990s, environmentalists had grand expectations that the insurance industry, as a powerful bloc of institutional investors, would change its investment patterns and move the global economy toward renewable energy, away from fossil fuels.[15] The idea that it might be possible to invest our way to a resolution to climate change seems to have caught on. Climate Wise, a consortium of insurance and reinsurance corporations that aims to include climate change in investment and risk assessments across the economy, is a flagship governance initiative from the insurance industry.[16] The Carbon Disclosure Project is an initiative that informs institutional investors of the carbon emissions of the companies they are investing in—over 2,000 companies reported their emissions in 2009.[17] Similarly, the Investor Network on Climate Risk is a $6 trillion network of investors that "promotes better understanding of the financial risks and opportunities posed by climate change."[18] Climate governance has gone corporate.[19]
- *The Cities for Climate Protection* (CCP) program of the International Council for Local Environmental Initiatives (ICLEI) coordinates actions of hundreds of municipalities (over 1,000 in 30 countries)[20] that pledge to work toward climate change mitigation through a common plan.[21] This network of municipalities has the potential to significantly contribute to a global response to climate change in and of itself, given its transnational nature and the fact that the network represents 15% of global carbon dioxide emissions.[22] As Betsill and Bulkeley argue, the cities program "has created its own arena of governance through the development of norms and rules for compliance with the goals and targets of the network."[23] The cities network also has an impact on efforts at other levels. Because cities are embedded in larger governmental structures, their efforts at promotion of climate protection contributes to climate politics at the national and multilateral levels and is now seen as a

key aspect of multilevel governance.[24] This subtler influence is not just encompassed by traditional lobbying and is, instead, an attempt "to reframe an issue which is usually considered in global terms within practices and institutions which are circumscribed as local."[25]

- *Carbon emissions trading systems* have emerged among coalitions of subnational actors. When nation-states refuse to move, subnational governments sometimes fill the vacuum.[26] One critical example of this phenomenon is how activist governors in the United States in partnership with provincial leaders in Canada have begun working to establish carbon markets that are simultaneously subnational and transnational. The mechanism of choice has often been emissions cap and trade systems where a group of actors agrees to cap emissions at a certain collective level and allocates permits for each member of the group to emit greenhouse gases. Those who emit less than their allocation can sell unneeded permits to those who emit more than their allocation. Over time, the cap on emissions ratchets down. Multiple cap and trade systems are in development through partnerships between U.S. states and Canadian provinces as well as a range of other actors.[27]

- *The Asia-Pacific Partnership for Clean Development and Climate Change and the Major Economies Forum on Energy and Climate*[28] are new multilateral initiatives. The stalemate in the ongoing negotiations over the Kyoto Protocol and its aftermath (hereafter the Kyoto process) has not stifled all multilateral approaches to climate governance. We have witnessed the emergence of these two initiatives, which are competing multilateral approaches. Both founded by the United States, they also arose through frustration with the Kyoto process, but with perhaps less progressive aims. Both of these experiments have a controversial relationship with the Kyoto process, and they offer a substantially different means of responding to climate change.[29] Whereas the Kyoto process is universal, these experiments are based on small-group negotiations with seven states in the Asian pact and 17 states participating in the Major Economies Initiative. Whereas the Kyoto Protocol is binding, these experiments stress voluntary measures. Whereas the Kyoto Protocol focuses on emissions reductions, these experiments turn to fostering technological innovation.[30]

Far from lacking a response to climate change as the UN process has floundered, the world is, rather, awash in different approaches. Dozens of climate governance experiments are shaping how individuals, communities, cities, counties, provinces, regions, corporations, and nation-states respond to climate change. The crucial task is understanding the significant opportunities and challenges for addressing climate change that arise as more and more actors engage in experimentation with perhaps the most significant governance challenge of the twenty-first century. Two motivating questions serve to guide this book toward such understanding.

First, why and how did experimentation emerge? While the circumstances surrounding the emergence of individual experiments are unique, the process and characteristics of experimentation are more general and can be apprehended. As the perceived legitimacy and effectiveness of the multilateral treaty-making process erodes, the world moves away from the conventional wisdom that the UN-led negotiating process is *the* way to address climate change. Consequently, we are transitioning into an era of experimentation, with enormous consequences for how the world responds to climate change. I examine how climate governance experiments both emerge from and contribute to the erosion of consensus around a megamultilateral response to climate change.

But so what? It is more important to consider the implications of experimentation—how climate governance experiments influence the shape and evolution of the global response to climate change. Climate governance experiments have the potential to drive agendas and mold the means of addressing climate change, in part determining whether we are likely to achieve an effective response to climate change. Therefore, I also examine the *impact* of experimentation on how climate governance is conceived and how it is likely to proceed.

Responses to climate change, both the traditional multilateral approach and the experiments chronicled in this book, have specific answers to some crucial questions:

- Who has a responsibility to respond to climate change?
- Who makes the rules that shape how communities will respond to climate change?
- What kinds of responses are appropriate for climate change?

The megamultilateral process has clear answers to these questions: nation-states are responsible, and they make the rules through the negotiation of legally binding international treaties. The initiatives examined in this book have different answers, and those answers have consequences for deciding which actors will lead the response to climate change and which types of responses will be seen as legitimate. By examining experimentation broadly—how experiments emerge and interact with each other and the ongoing multilateral process—I uncover the kind of climate governance currently emerging in the void left by stalemate in the multilateral treaty-making process. Will a coherent system of governance emerge from disparate initiatives? Will it supplant or complement treaty-making as the mode of a global response to climate change? These questions have enormous practical implications for determining how the world responds to climate change and whether experimentation is leading the world into a void of ineffectual responses.

The rest of this chapter sets the stage for telling the story of climate governance experiments. Background on the challenge of climate change and the multilateral response to this challenge provides a foundation for thinking about the two trends that together form the central dilemma for this book: the concurrent demise of effective megamultilateral treaty-making and the rapid emergence, but uncertain effectiveness, of new initiatives. I then introduce the initiatives that constitute the world of climate governance experiments before briefly outlining the course of the chapters to follow.

The Challenge of Climate Change and the Megamultilateral Response

Though the climate system is almost unbelievably complex, the logic of the climate change problem is relatively simple to describe. The earth's atmosphere acts as a greenhouse whereby various gases (carbon dioxide, methane, chlorofluorocarbons, water vapor, and others) absorb solar radiation that would otherwise be reflected back into space from the Earth. This greenhouse effect itself is beneficial as it keeps the planet warm and allows life to flourish in the forms with which we are familiar. However, since the Industrial Revolution, humanity has been emitting more and more greenhouse gases (e.g., carbon dioxide, methane, chlorofluorocarbons), mostly through the burning of fossil fuels, increasing their concentrations in the atmosphere and thus increasing the warming effect. Potential effects of increased warming include sea level rise, increases in the frequency and severity of storms and droughts, changed precipitation patterns, altered disease vectors and trajectories, species migration, reduced agricultural productivity, and more.

Yet this simple description of how global warming takes place belies the enormous complexity of devising means of mitigating and/or adapting to the consequences of climate change. The climate is one of the most complex systems humanity has ever attempted to understand, and while there is broad consensus that climate change is taking place because of human activity, there are a number of uncertainties and complicating factors that restrict our ability to make straightforward policies to address the problem. Scientific challenges include comprehending and tracing:

- The exact relationships between concentrations of greenhouse gasses and temperature increases on the one hand and climatic changes like increased severity and frequency of storms, cycles of droughts and floods, and patterns of precipitation on the other

- How natural variability in the climate can mask and/or exacerbate the effect of anthropogenic greenhouse gas emissions
- The uncertain magnitude and geographically variable nature of the effects of climate change

In addition to scientific challenges, and in some cases directly related to the scientific uncertainties, the social/political obstacles to addressing climate change can hardly be overestimated:

- *Greenhouse emissions arise from virtually every human activity.* Most current industrial and agricultural processes produce greenhouse gases. The world's economy significantly relies on fossil fuel use.
- *Dependence on fossil fuels is uneven.* While the global economy runs on fossil fuels, there is disparity among both consumers and producers of fossil fuels— in other words, some countries produce a lot of fossil fuels, others consume a lot of fossil fuels, and many that consume less would like to consume more.
- *Per capita greenhouse gas emissions vary significantly.* While absolute emissions from India and China rival those found in the United States and EU, the per capita emissions are wildly divergent. According to the International Energy Agency, in 2007 the average person in the United States produced over 19 tons of carbon dioxide, while the average person in India and China produces 1.2 and 4.6 tons, respectively.[31]
- *Protecting the climate promises diffuse benefits in the future, while engendering concentrated costs now.* Put simply, it is difficult to generate political will, especially across political jurisdictions, to solve a problem when identifiable groups must pay up front to generate benefits for the whole world sometime in the future. Scientists agree that the world must take action now to change the nature of our world's economy and wean ourselves away from fossil fuels so that decades or even a century in the future, our climate remains hospitable for our grandchildren. This creates an enormous incentive to delay and significantly hampers efforts to generate urgent action in the present.

These challenges make climate change a "super wicked" problem that consistently and chronically poses virtually insurmountable challenges to formulating traditional, rational solutions.[32]

Characteristically of wickedness, defining the "problem" of climate change is an inherently political exercise. Is it an economic problem or an environmental problem? Is it a problem of mitigation or adaptation? Is it a problem of underdevelopment or overdevelopment? Answers to these questions determine how policies are developed, and they cannot be found in climate science. Compounding this definitional challenge is the fact that solutions to wicked problems like climate change are hard to come by and ultimately uncertain. It is impossible to consider every solution to the problem or fully evaluate the solutions that are chosen because of the scientific and political uncertainties mentioned above. Further, given the time scales involved with climate change, even when solutions are chosen, we will not know if we chose correctly for decades.

The kicker is that wicked problems are inherently path dependent.[33] The definitions and solutions ultimately settled on, however imperfectly, constrain what we consider possible in the future because implementing potential resolutions to super wicked problems "leaves traces that cannot be undone."[34] In the latter 1980s, the international community began to design and implement a megamultilateral resolution to climate change. This initial understanding had consequences, conditioning how we conceive of climate change and its potential solutions to the point that it is a cliché to say that climate change is a global problem that requires global solutions.[35] The world's choice of response (multilateral treaty-making) has determined our very understanding of the problem we face (a distinct focus on the global nature of the problem instead of the varied local sources and effects of climate change).

In light of the challenges, the attempt to govern climate change—to decide on the rules and institutions necessary for ultimately decarbonizing economies and societies on a global scale and dealing with the effects of climate change that we are already destined to endure—has been a fascinating, if disappointing, two-decade exercise. Fascinating because it has been an effort at international collective action on a grand scale—nearly 200 nation-states have been engaged in negotiations since 1990 attempting to devise international agreements to resolve the problem. Disappointing because the international community has struggled mostly in vain to overcome the obstacles to collective action and devise an effective multilateral response to climate change. The early stages of climate governance were aided by a dominant definition of the problem shared throughout the international community: climate change was a transnational *environmental* problem (i.e., mitigating greenhouse gas emissions was the issue) that required multilateral treaty-making through a negotiation process that included all nation-states.[36] Acting on this common knowledge reinforced the idea that megamultilateralism was *the* way to address climate change. With few exceptions, NGOs oriented their activities toward convincing nation-states to take action.[37] Cities and provinces had very little in the way of climate policy.[38] Individuals looked to their nation-states to take action and urged either restraint or boldness depending on how they understood the urgency of the problem. While the content of international treaties was subject to considerable debate, the megamultilateral model of governance permeated actors' definitions of the problem and practices across political levels. The megamultilateral response developed in three stages.

STAGE 1: THE UN FRAMEWORK CONVENTION
ON CLIMATE CHANGE

Climate change leapt onto the global political agenda in the late 1980s with the development of significant scientific evidence that the globe was warming with uncertain, but likely adverse, consequences.[39] Some countries were motivated to take significant action almost immediately (Europe, small island nations).[40] Others were not convinced (oil producers, large developing countries, the United States, Canada, Australia, Japan, Russia), and some worked to slow the response. Initial negotiations culminated in the signing of the UNFCCC at the Rio Earth Summit in 1992. At the very outset, the international community's response to climate change was megamultilateral, with over 100 states attending each of the negotiating sessions leading to the 1992 Earth Summit.

This initial set of negotiations came fast on the heels of perhaps the biggest success story in multilateral environmental agreements—the treaties that addressed the problem of ozone depletion: the 1987 Montreal Protocol on Substances that Deplete the Ozone Layer (hereafter Montreal Protocol) and the 1990 London Amendments to the Montreal Protocol. Flush with the success of agreeing to essentially phase out ozone depleting chemicals with these treaties, the international community turned its attention to climate change. While the form of the response would mirror the governance mechanism employed to address ozone depletion (universal multilateral negotiations toward a binding treaty, an initial framework convention to be followed by specific protocols, a commitment to a differentiated response to the problem based on development levels), climate change would prove to be a much more complex and difficult problem to confront.

The United States was the main obstacle to quick action, forswearing any moves to include binding greenhouse gas emissions reduction targets in an initial framework convention for climate change that would lay out the process of addressing the problem. The Europeans and small island nations pushed for decisive and forward-looking action, but were ultimately thwarted. There were rifts along North-North (U.S.-EU) and North-South lines. Southern states, bolstered by the precedent set in the Montreal Protocol, urged Northern states to take the first significant actions to address climate change. The United States and the EU jousted over binding emissions, and the United States also pressured Southern states to take an active and immediate role in the fight against climate change. The result of these debates was a relatively weak framework convention. It merely mandated that Northern states compose national emissions inventories and actions plans. There was a broad and voluntary commitment to stabilize the climate and to return to 1990 emission levels. Overall, the UNFCCC reinforced the notion of common but differentiated responsibilities—a two-track response to climate change based on level of development—but did little more than set general aspirational goals.[41] In this way it was quite a close parallel to the weak 1985 Vienna Convention to combat ozone depletion, which preceded and set the stage for the more auspicious Montreal Protocol. Thus while the UNFCCC was weak, there was some optimism that a similarly weak framework convention could produce a similarly effective protocol in the years to come.

STAGE 2: THE KYOTO PROTOCOL

This hope did not come to fruition. The second set of negotiations began in earnest in the mid-1990s and culminated in the 1997 Kyoto Protocol to the UNFCCC. The United States changed course in 1995, agreeing for the first time to undertake binding emissions reductions, raising the prospects for the negotiation of an effective protocol to the UNFCCC. Unfortunately, debates that slowed and weakened the UNFCCC were never resolved and proved difficult to overcome in this second set of negotiations as well. North-South differences played a critical role as the U.S. Senate, in the form of the Byrd-Hegel resolution, a pledge never to ratify a climate agreement that failed to bind developing countries with similar restrictions agreed to by the United States. In addition, while the United States agreed in principle to emissions reductions, there was significant debate with the European Union over how deep the cuts would be and how reductions could be accomplished. The United States pushed vigorously for "flexible mechanisms" that would allow nation-states to pursue emissions reductions where they were cheapest, and the Europeans favored more domestically oriented reductions.

In contrast to 1992, however, compromise was achieved at Kyoto. While the Kyoto Protocol is routinely criticized, it did realize modest accomplishments. It requires emission reductions from Northern states, an average of 5% below 1990 emission levels by 2008–2012. It includes flexible implementation measures aimed at keeping the costs of emissions reduction relatively low. Finally, it contains an initial foray into the delicate issue of the North-South impasse, with the clean development mechanism (CDM), which allowed Northern states to get credit for emissions reductions by paying for greenhouse gas reduction projects in Southern countries.

The process of bringing the Kyoto Protocol into force (with the ratification of at least 55 countries that are responsible for at least 55% of greenhouse gas emissions from among developed countries) and finalizing its details engaged the international community from 1998 through 2005. Each year from 1998 to 2001, parties to the UNFCCC and signatories of the Kyoto Protocol met to operationalize it, coming to a final agreement with the Marrakech Accords of 2001. Looming over this process, however, was continued

U.S. recalcitrance. Working from within the shadow of the 1997 Byrd-Hegel resolution, the United States attempted to persuade Southern states of the need to take on comparable commitments. This effort failed, and in 2001, citing uncertainty and unfairness, the Bush administration withdrew the U.S. signature from the Kyoto Protocol. The rest of the world moved forward, bringing the Kyoto Protocol into force in February 2005 with Russia's ratification. Even though those states that ratified the Kyoto Protocol will meet their goal of a collective 5% reduction,[42] the only binding international treaty for addressing climate change moved forward into an uncertain future because it failed to significantly engage the United States or the large industrializing countries.

STAGE 3: DEVELOPMENT OF IMPASSE AND DEMISE OF MEGAMULTILATERALISM?

Overlapping the push to bring the Kyoto Protocol online was the third stage of the megamultilateral response—the development of stalemate. In 2001, a difficult collective action problem worsened with the U.S. withdrawal from the Kyoto Protocol. Political economy scholars explain this by referring to economic interests.[43] Given its preeminent position as an energy consumer and carbon dioxide producer, the United States does not want to incur what would be significant costs to its economy to deal with the problem, especially in the absence of action by major economic competitors like China. Large developing countries that have rapidly grown in terms of energy consumption and carbon dioxide emissions (in absolute if not per capita terms) prioritize development over action on climate change and also argue that a problem historically caused in the North should be dealt with by Northern states first. The United States is reluctant, at best, to take significant action. China, India, Brazil, and other developing states are reluctant, at best, to take significant action. The European Union which has taken significant action,[44] has not been able to convince either side to make significant concessions. Impasse.

The most optimistic analysts are worried that progress in climate governance has been too slow.[45] Less generous observers have harsher words for the UN process. Depledge soberly observes that the Kyoto process "has not only got 'stuck' but is digging itself into ever deeper holes of rancorous relationships, stagnating issues, and stifling debates" and that this "ossifying" negotiating approach and governance framework cannot produce the rules and innovations necessary to meet the climate challenge.[46] The United States, especially, has not been inclined to provide leadership, given its dependence on fossil fuels. Large developing countries have also consistently signaled that they prioritize development goals over climate protection.

Thus a stalemate has developed whereby European and small island nations want to move forward quickly. The United States (and until 2007 Australia) pulled out of the Kyoto Protocol process. India and China and other large Southern nations refuse to move until Northern states take bolder actions (though some signs of movement in this position were evident at the Copenhagen negotiations in 2009). Canada, Russia, and Japan are wary of further commitments, and Canada explicitly will not meet its Kyoto Protocol obligations. This has been the consistent story since the Kyoto Protocol was signed. The Europeans and major Southern states push for significant actions by Northern states, and the United States, and to a lesser extent Japan, Russia, and Canada, work to both reduce and slow the response to climate change and push for concomitant Southern actions. It would seem that the megamultilateral response is slowing if not grinding to a halt.

The fact that the criterion for "success" in the 2007 UNFCCC COP negotiations in Bali consisted of agreeing to the Bali Roadmap, which called on nation-states to negotiate a post–Kyoto Protocol agreement essentially *from scratch* by 2009, did not bode well for this approach to climate governance.[47] At the 2008 UNFCCC COP negotiations in Poznan, the European Union began to pull back from some of its more stringent goals in the face of the global financial crisis.[48] Most recently, hopeful signs coming from the United States since the election of Barack Obama and from China in terms of willingness to consider significant action raised expectations for the Copenhagen negotiations. Yet even with these advantages, the international community was not able to achieve a legally binding replacement for the Kyoto Protocol. In 20 years of trying, the multilateral treaty-making process has achieved the Copenhagen Accord, a political agreement that fails to go significantly beyond the original UNFCCC and that is fearfully inadequate for meeting the challenge of climate change.

The two-decade story of climate negotiations recounted above is remarkable for how consistent it is. For the vast majority of the last 20 years, there was no significant questioning of the basic assumption that brought us the UNFCCC and the Kyoto Protocol—the appropriateness and efficacy of megamultilateralism. Much of the governance reform literature centers on how to do multilateralism (mega or otherwise) better.[49] Critics decry the lack of progress and criticize the results of the negotiations, but until very recently there have been too few serious proposals to approach climate governance differently. Ironically, however, the Kyoto Protocol may mark *both* the high point of universal, multilateral mechanisms for addressing global climate change *and* the onset of the demise of this mechanism. For all the efforts of negotiators and urgency surrounding this issue, multilateral treaty-making has consistently failed to produce treaties and agreements that effectively address climate change. It may be time to concede that there is a mismatch between this type of treaty-making and the problem of climate change; that global treaty-making, as attempted in the last two decades, cannot catalyze the societal and economic transformation necessary to avoid the potentially catastrophic consequences of climate change.[50] The failure at Copenhagen should be viewed as an opportunity for thinking differently about the way climate change is governed.

Climate Governance Experimentation

If Copenhagen does end up representing a major turning point in climate governance it will be partially due to the activities that have emerged in the face of stalemate at the interstate level. We are no longer witnessing a singular (faltering) global response to climate change; we are instead observing multiple global responses. This book is an attempt to make sense of these multiple global responses, to understand how and why they arose, if and how they are organized, and the impact they might have.

Accomplishing these goals begins with the (not so) simple task of clearly demarcating the arena of study—identifying climate governance experiments. The range of activities seeking to fill the void of a breakdown in the multilateral treaty-making process is wide, but the analysis in this book is restricted to initiatives that meet relatively strict criteria designed to capture innovations of a particular sort—those that are *experimental* attempts at *governance*. The term "governance" has grown in popularity in recent decades, and with popularity has come a proliferation of meanings.[51] Yet at its core "governance" implies steering or rule-making outside traditional channels of centralized authority.[52] As James Rosenau says, governance is a matter of having the capacity to get things done, without the formal authority to do so.[53] Governance is about making rules above, below, and between established political authorities. Experimentation implies innovation and trial and error (rather than the more formal controlled laboratory definition) with new forms of governance unrelated or only loosely connected to the traditional mechanism of multilateral treaty negotiations.

With this in mind, in order to count as a climate governance experiment for this book, a climate related initiative must meet three criteria.[54] First, climate governance experiments are primarily engaged in *explicitly* making *rules* that shape how communities respond to climate change. I did not begin with a set of a priori criteria for what actually counts as addressing climate change—uncovering different framings of this issue is precisely the point. However, I only include initiatives profess and could be reasonably considered to have addressing climate change as a primary objective. Further, I only include initiatives with a conscious intention to create/shape/alter behavior by setting up rules (broadly conceived as including principles, norms, standards, and practices) for a community of implementers (whoever and whatever they may be) to follow. For this book, intentionality and potential authority over a community is a key marker of governance experimentation.

Second, climate governance experiments are independent from the Kyoto process or national regulatory measures. Initiatives that are designed to aid states in meeting Kyoto goals or national regulations are not experiments. This sorting rule is crucial so as to capture innovations that are outside the traditional megamultilateral process. Individual nation-states have come up with an interesting array of climate initiatives designed to either make the Kyoto process work or to enhance national climate action.[55] These are crucial aspects of the global response to climate change. They are not, however, experimental for this research. I do count as experimental the new multilateral initiatives that have emerged among small groups of nation-states that are aimed at technological cooperation rather than treaty-making. Applying an old tool (multilateral cooperation) to a new problem (climate change) with a different goal (not making a legally binding treaty) is experimentation.

Third, climate governance experiments cross jurisdictional boundaries of some sort. I only include initiatives designed to function across jurisdictions, whether vertically (local-regional-national-transnational) or horizontally (networks of similar actors across boundaries). This third classificatory rule significantly constrains the population of experiments, ruling out, for example, many municipal climate action plans. Thousands of cities have developed climate action plans. This is innovative because climate policy has generally been the purview of national governments. However, for both practical and theoretical reasons, individual municipal initiatives (and other single jurisdiction initiatives) are not included.

This is a practical matter because without this rule, capturing any reasonably coherent picture of experimentation would be nearly impossible—there would simply be too many possible experiments. This limiting rule is important substantively as well. While taking on climate change is innovative for cities, these actors have well-developed means for making rules—for governing. Experimentation is, rather, a process of making rules outside well-established channels.[56] While individual city initiatives are not included, networks of cities like the CCP program do count as climate governance experiments. I am most interested in initiatives that have to make rules from whole cloth. For example, there is no established institution for U.S. states and Canadian provinces to cooperate and make climate agreements, yet the Western Climate Initiative (WCI) and the Midwestern Greenhouse Gas Reduction Accord have been forged through the cooperation of these actors across a national border and without the input or facilitation of the national governments of Canada and the United States.

With this operational definition my graduate research assistant and I identified 58 experiments (see Table 1.1) meeting the three criteria through a systematic search of side events at the UNFCCC COPS from 20036-2008 (COP 9-14),[57]

Table 1.1 **Climate Governance Experiments***

2degrees	Social networking platform for actors working in corporate environmental sustainability, climate change and green technologies
Alliance for Resilient Cities	An Ontario-based network of municipalities focused on adaptation to climate change.
American Carbon Registry	Apparently the world's first private emissions registry started in 1997 as the Greenhouse Gas Registry.
American College & University Presidents Climate Commitment	Pledge and program to eliminate greenhouse gas emissions at US colleges and universities.
Asia Pacific Partnership on Clean Development and Climate	Voluntary partnership among select countries to cooperate on technological development and implementation in a number of sectors.
Australia's Bilateral Climate Change Partnerships	Agreements and partnerships signed between Australia and other countries (developed and developing) to take action on climate change.
Business Council on Climate Change	Partnership of San Francisco Bay Area businesses committed to reducing their green house gas emissions.
C40 Cities Climate Leadership Group	A network of the world's largest cities created to share best practices and develop collaborative initiatives to do with city-specific issues.
California Climate Action Registry	Voluntary greenhouse gas registry now operating under the Climate Action Reserve.
Carbon Disclosure Project	Resource/database for institutional investors to inform their investment choices based on emissions reported by the world's largest corporations.
Carbon Finance Capacity Building Programme	Partnership to encourage the use of carbon finance to address greenhouse gas emissions in mega cities of the South.

continued

Table 1.1 (continued)

Carbon Rationing Action Groups	Network of local groups to support and encourage one another in reducing individual carbon footprints.
Carbon Sequestration Leadership Forum	Framework agreement between governments to promote and develop carbon capture and storage technology.
CarbonFix	Both a labeling standard setter and greenhouse gas offset supplier.
Chicago Climate Exchange	Private cap and trade system whose members make a legally binding emission reduction commitment.
Climate Alliance of European Cities with Indigenous Rainforest Peoples	Association of European cities and municipalities that have entered into a partnership with indigenous rainforest peoples.
Climate, Community, and Biodiversity Alliance	Partnership of international NGOs that develops and manages standards to promote the delivery of social and environmental benefits from land-based emissions reduction activities.
Climate Neutral Network	A web-based platform for networking and the sharing of best practices on reducing and offsetting greenhouse gas emissions.
Climate Savers	Corporate partnership between major corporations, organized by the WWF, to increase efficiency in operations/products to voluntarily reduce their greenhouse gas emissions.
Climate Wise	An association of insurance-related companies/organizations established to collaborate on climate issues.
Clinton Climate Initiative	Program of the Clinton Foundation that seeks to provide direct assistance to individual cities and facilitate the sharing of best practices.
Community Carbon Reduction Project	A UK-based network of local community partners that engage in education, research, and outreach to cut their CO2 emissions to meet a target of 60% reduction by 2025.

continued

Table 1.1 (continued)

Conference of New England Governors and Eastern Canadian Premiers Climate Change Action Plan	Voluntary agreement to pursue coordinated actions on climate change in the region.
Connected Urban Development Programme	Partnership between Cisco and cities to create urban communications infrastructures to reduce carbon emissions.
Cool Counties Climate Stabilization Initiative	Network of counties created to address climate change.
Covenant of Mayors	Commitment by European cities to go beyond the EU CO2 emissions objectives through energy efficiency and clean energy initiatives.
e8 Network of Expertise for the Global Environment	Non-profit international group of nine major electricity companies from G8 countries, promoting sustainable development through electricity sector projects and capacity building activities worldwide.
Edenbee	Web-based social network (a la facebook) designed to encourage users to reduce their carbon footprints.
EUROCITIES Declaration on Climate Change	Agreement by European cities to fight climate change at the local level
Evangelical Climate Initiative	An agreement by evangelical leaders to motivate their followers to protect the climate.
Global Greenhouse Gas Register	A global, corporate-wide emissions registry for companies based in developing or other countries (i.e., US) not subject to Kyoto Protocol obligations (now defunct).
ICLEI Cities for Climate Protection Campaign	Campaign that seeks to promote the development and implementation of greenhouse gas emission reduction strategies among local and municipal governments.

continued

Table 1.1 (continued)

Institutional Investors Group on Climate Change	A forum for collaboration between pension funds and other institutional investors in Europe to address the investment risks and opportunities associated with climate change.
International Climate Action Partnership	International forum of governments and public authorities that are engaged in the process of designing or implementing carbon markets.
Investors Group on Climate Change	Collaboration of Australian and New Zealand investors focused climate-related investment risk.
Investor Network on Climate Risk	A $7 trillion network of investors geared toward integrating climate risks into investment decisions.
Klimatkommunerna	Association of Swedish municipalities, counties and regions working actively on climate issues.
Major Economies Forum on Energy and Climate	A "complementary" process to Kyoto framework that brings together leaders of the world's 17 largest economies to discuss climate.
Memoranda of Understanding on Climate Change initiated by the State of California	Memoranda between California and various nation states and sub-national units/states/provinces for joint efforts on climate change.
Methane to Markets	A framework agreement between countries to promote methane recovery internationally.
Midwestern Greenhouse Gas Reduction Accord	Policy framework to develop a cap-and-trade mechanism amongst Midwestern US states and Canadian provinces.
National Association of Counties County Climate Protection Program	Project to provide counties with best practices, tools and resources to assist them in developing and implementing successful climate change programs.

continued

Table 1.1 (continued)

Network of Regional Governments for Sustainable Development	International network of sub-national regional governments based on partnerships and bilateral cooperation agreements among members.
North South Climate Change Network	Network of an NGO, university, and local communities with the goal of improving the Ontario's knowledge of, and response to climate change.
Ontario-Quebec Provincial Cap-and-Trade Initiative	Inter-provincial cap-and-trade program (now folded into the Western Climate Initiative)
Regional Greenhouse Gas Initiative	Regional cap and trade program among Northeast and Mid-Atlantic states in the US.
Renewable Energy and Energy Efficiency Partnership	International NGO that works in partnership with business, civil society, and government actors to reduce the barriers to the uptake of renewable energy and energy efficiency technologies and projects.
Southwest Climate Change Initiative	Framework agreements between two Southwest US states to coordinate emissions reductions.
The Climate Group	An independent, nonprofit organization dedicated to advancing business and government leadership on climate change.
The Climate Registry	A collaboration between U.S. states, Mexican and Canadian provinces, and Native American organizations aimed at developing and managing a consistent North American greenhouse gas emissions reporting system
Transition Towns	Network/set of principles that encourages communities to "relocalize" all essential elements that the community needs to sustain itself

continued

Table 1.1 (continued)

Union of the Baltic Cities Resolution on Climate Change	Resolution among Baltic cities to combat climate change and make plans for adaptation
US-China Memorandum of Understanding to Enhance Cooperation on Climate Change, Energy and the Environment	Bilateral Climate Change MOU between the US and China.
U.S. Mayors Climate Protection Agreement	Agreement by US Conference of Mayors to advance the goals of the Kyoto Protocol in the US.
UK Bilateral Climate Change Agreements with US States	Formal agreements between UK government and various US states
West Coast Governors' Global Warming Initiative	Collaboration between three Western US states that produced a set of recommendations on cooperative strategies.
Western Climate Initiative	Network of states and provinces in the United States, Canada, Mexico to cooperate on climate action and a regional cap and trade system.
World Business Council for Sustainable Development	CEO-led, global association geared toward sharing of best practices and knowledge on climate change.

*See Appendix for a listing of websites from which information on the experiments was drawn.

and news reports gathered from the Dow Jones Interactive database from 1990-2008. The side events are generally populated by a variety of nonstate actors discussing and publicizing their climate activities, and as such, are ideal sources for identifying experimental initatives. The media reports were an instrumental source for tracking the emergence of experiments back to the initiation of climate change as an international policy issue. As interest in climate change has increased in the last two decades, the media has taken to reporting not just the international negotiations but also initiatives beyond the multilateral process. In addition to these systematic searches, a number of potential experiments were suggested by colleagues and experts in the field of climate policy.

This data-gathering strategy has some limitations. I do not claim to have gathered either the entire population or a representative sample of climate governance initiatives. The search was bounded by language (English), access to the UNFCCC process, and the scope of available media. Indeed, ultimately, the dataset cannot form a representative sample because the full extent of the population of experiments is unknown. Yet the diversity and dynamism of experimental climate initiatives captured in Table 1.1 is fascinating in and of itself and this dataset does provide a solid foundation for rethinking climate governance, even if it is not a random, representative sample or full population of experimental initiatives.[58] Examining a relatively large set of diverse climate governance initiatives, analyzing how they emerged, are organized, and influence the global response to climate change provides a window on a key set of climate governance dynamics beyond the multilateral process and illuminates how climate governance may evolve even if it is not a perfectly representative sample of the range of climate governance activities.

Indeed, the recent burst of innovation captured in Table 1.1 appears to be something different, something of potentially enormous but uncertain consequence. Yet our understanding of these experiments, especially as a collection of initiatives, is still rudimentary. Scholars do recognize that "the only appropriate response [to climate change] is a multilevel governance response in which concurrent policy processes at all levels identify policy space and foster initiatives as well as put pressure on the other governance levels."[59] *Individual* climate change initiatives that go beyond the UN process have begun to garner significant attention in academic writing and popular media,[60] but too few pursue understanding of the current flurry of activity comprehensively examining the relationships between and organization of these new initiatives.[61] This book attempts to bring a modicum of order to observations of the cacophony of climate initiatives. It provides a holistic view, considering that the diverse experiments in Table 1.1 may be of a kind, related by the process of experimentation and participating (consciously or unconsciously) in nascent experimental system of governance. It provides a way to think differently about climate governance and the implications of new initiatives in the global response to climate change.

The Path into the Void

Chapter 2 begins the process by opening a window onto the world of climate governance experiments, examining not just who is engaged in experimentation but, more important, what experiments do. The first half of the chapter describes the climate governance experiments from Table 1.1 laying the foundation for an analysis of the characteristics of the collection of experiments. This second half reveals that experiments are bound together by a common liberal environmental ethos that stresses the compatibility of economic growth and environmental protection.[62] However, in addition to this similar philosophical foundation, experiments are functionally differentiated. There are a limited number of distinct modes of responding to climate change in the experimental world: networking, infrastructure building, voluntary action, and accountable action.

Chapter 3 posits a way to make sense of the emergence and impact of experimentation. The climate governance experiments listed in table 1.1 are both symptoms and drivers of the transition away from a sole focus on multilateral treaty-making in the global response to climate change. I argue that feedback between actors (nation-states, cities, provinces, corporations, individual people) and what I call the governance context (ideas about what the appropriate response to climate change should be) catalyzes the change we are seeing and is prompting the organization of a coherent, if still nascent, experimental *system* of climate governance.

Chapters 4–6 examine in detail the kind of response that is emerging in the world of climate governance experimentation. Chapter 4 explores canonical examples of experiments in each of the four modes of experimental governance in some depth. These vignettes serve to check the plausibility of the story presented in chapter 3 and to explore how the experimental system is structured by the functional differentiation observed in chapter 2. Chapters 5 and 6 analyze the organization of the experimental system by focusing on clusters of experimental activity. Chapter 5 concentrates on the emergence of programs to develop and deploy climate-friendly technology in municipal networks. Chapter 6 explores the development of carbon markets.

Chapter 7 considers the potential broader significance of experimentation and the nature of the ongoing global response to climate change. In 1988, at one of the first major global warming conferences, scientists declared that "humanity is conducting an unintended, uncontrolled, globally pervasive experiment" that began at the dawn of the Industrial Revolution: emitting greenhouse gases and waiting to see the effects.[63] This grand experiment is now being met with climate governance experiments. Whether we can address the climate experiment effectively with governance experiments alone or in combination with multilateral treaty-making remains to be seen. In the concluding chapter I reflect on the potential to be found in climate governance experimentation and the research and work that remains to be done to improve the global response to climate change.

2

The World of Climate Governance Experimentation

Climate change is everywhere we look—not necessarily the physical phenomenon itself, though evidence that climate change may be worse and occurring more rapidly than earlier thought is now coming to light,[1] but climate change is becoming a ubiquitous lens through which we view our world. Consider the weather. Now any conceivably anomalous weather event is attributed to climate change or touted as a refutation of it. Episodes in 2010 like the brutal Russian heat wave and horrific flooding in Pakistan or the aftermath of Hurricane Katrina in 2005 spur significant concern for climate action.[2] Noted climate skeptic Senator James Inhofe built an igloo near the Washington Memorial in Washington, D.C., during an unusual winter storm in 2010 to publicize his contention that climate change is a hoax.[3] The point is not to say that both are equally misguided—they are not. In fact, anomalous weather events are thought to be more likely because of climate change (both powerful hurricanes and weird snowstorms) even if we cannot attribute any specific event to climate change.[4] The point is, rather, the ubiquity of the discussion of climate change and the way it informs and molds thinking about the world—for both skeptics and those convinced by the scientific evidence.[5] In the important book *Why We Disagree about Climate Change*, Mike Hulme in fact argues that climate change is no longer a problem to be solved but a condition of modern life—it is the context within which we live our lives.[6]

But the ubiquity of climate change has yet to be reflected in successful treaty-making or significant national responses throughout the international community. It is perhaps not surprising, then, that experimental activity, both by my definition and in the colloquial sense,[7] is now a significant aspect of the climate change issue area. As climate change comes to define the context of modern life, it is becoming infused into multiple aspects of daily life. It is impossible to avoid advertisements for carbon offsets when traveling, or to elude calls for replacing incandescent light bulbs with compact fluorescents. The media has chronicled the surge of fuel efficient and hybrid vehicles that are leading the attempted resurrection of the auto industry. Thousands of cities are implementing local climate action plans. Cap and trade has become an increasingly commonplace part of the public lexicon as subnational governments and nation states debate this response to climate change. Climate governance experiments are of interest in this context both because of what they are—innovations in approaching this problem—and what they *may* be able to do: catalyze a more effective global response to climate change.

Climate governance experiments *are* initiatives that exist at the seams of established modes of shaping human behavior and are thus hard to categorize. They are not traditional political authorities—governments that have the legal authority to make binding rules. Even the experiments initiated by governments or that include governments as implementing actors, like the regional emissions trading initiatives among U.S. states and Canadian provinces, are not governments themselves. Experiment initiators act collectively and voluntarily without traditional legal governmental authority, even when, in certain cases, their members' governmental status is advantageous in carrying out experimental initiatives. Experiments are also not radically decentralized markets—though many aim to alter consumption and production patterns. Experiments like The Climate Group aim to harness market forces to address climate change, but they do establish rules of behavior rather than letting the blind hand of the market determine behavior. Further experiments are simultaneously local and transnational. The C40 network of global cities is a transnational movement generated by local municipal actions.

What climate governance experiments *may be able to do* for the global response to climate change is provide, first, a source of friction in politics and markets that catalyzes demand for broad transformation in societies and economies, and second, a source of smoothing that provides the technologies and institutions to respond to the demand for transformation. Experiments thus have the *potential* to create disruptive change, shaping how groups of individuals, corporations, cities, subnational governments, and even nation-states address climate change and altering the way politics and markets function.[8]

Climate governance experiments generate friction by pushing the boundaries of traditional notions of which actors are responsible for making rules, creating uneven sets of rules that actors must follow, and generating new coalitions committed to climate action. This friction, in turn, may be able to catalyze broader change as other actors (including national governments) and processes (e.g., multilateral negotiations) and arenas (e.g., global markets) react to the friction. Experiments can facilitate a positive reaction as well, smoothing the path of the response to climate change by building institutional, technological, and political capacity. It seems self-evident yet important to note that the experiments discussed here are experiment*ing*—they are having a go at a multitude of ways to attack this problem, learning lessons that may be more widely applicable. Through both of these processes, friction generation and smoothing, experimentation may be able to create significant momentum for the global response.

Exploring this conjecture about potential impact and assessing the possibilities and challenges to be found in climate governance experimentation requires a prior descriptive endeavor. In this chapter I examine who is participating in climate experimentation, where it is occurring, and what experiments are actually doing with the goal of sketching the contours of the collection of climate governance experiments. Data was gathered from organizational websites and documents to characterize each experiment by date of initiation, geographic location, the initiating and implementing actors, and the activities undertaken. Semi-structured interviews with representatives from initiatives augmented this data, helping to clarify what functions experiments perform and to uncover the how climate governance experiments are organized. The mapping of the experimental world that results provides the foundation for the efforts to make sense of experimentation and its implications in the chapters that follow.

Demographics of Experimentation

The first quality of the experimental world that stands out is how recent experimentation is. While climate change has been the focus of international governance efforts since the late 1980s, experimentation was slow to develop. Figure 2.1 reports on the temporal sequence of experiment initiation displaying the number of climate governance experiments that were initiated in each year from 1990 to 2009. The first half of the 1990s saw only four experiments initiated. These early ones—the World Business Council on Sustainable Development (1990), the Climate Alliance (1990), the e8 Network (1992), and the CCP (1993)— emerged during the negotiation of the UNFCCC and aftermath of its signing, but they were rare exceptions. In addition, in these early years, programs were more concerned with implementing multilateral initiatives or dealing with rainforest destruction and energy efficiency than striking out on their own independent responses to climate change.[9] The period 2001–2002 was a watershed for both the multilateral process and experimentation. While only 5 experiments emerged in these two years, experimental activity in subsequent years exploded. Of the 58 experiments in the database, 46 were initiated after 2001–2002, perhaps the nadir for the multilateral process in the aftermath of the American withdrawal.

The geographic scope of the experimental world is simultaneously limited and expansive. If we consider where experiments are initiated—the physical location of the entrepreneurial actors that create experiments—then it is evident that the experimentation captured in this research is mainly restricted to the developed world, mostly North America and Europe. Of the 58 experiments, 51 were

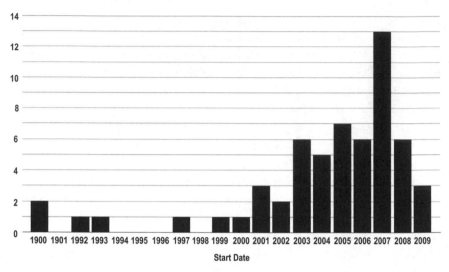

Figure 2.1 Experiment Initiation

initiated by actors in the global North,[10] with the remaining 7 initiated by a combination of actors in the North and South. None of the 58 experiments identified were initiated in the South alone.[11] Yet when we examine the scope of experiments in terms of the location of their activities (fig. 2.2), we find that the experimental world is expansive, with experiments functioning in the North and the South at local, regional, national, and global scales.

The notion of where experimentation is happening encompasses more than the physical location of initiators and activities—"where" also has an important political dimension. Examining the distribution of *types* of actors that initiate and implement climate experiments reveals striking diversity in the experimental world. Actors at multiple levels of political organization have decided to experiment with climate governance, from small groups of individuals through networks of the largest state economies on the planet. Figure 2.3 is a visual representation of this diversity. To be sure, this figure is not weighted by size of initiative, thus potentially problematically equating the Major Economies Forum of the 20 largest nation-state economies with Edenbee, a social networking experiment with just 4,500 members in its online community in the same graphic. Yet the diversity is interesting in and of itself and a key point is that almost any actor can conceive of being a player in climate governance and seeking to influence the response to climate change.

There is simply no single idea as to which actors should initiate or be responsible for carrying out responses to climate change in the experimental world. Further, there are significant attempts at crosstype governing—one kind of actor making experimental rules designed to be implemented by different types of actors. A slight majority of experiments have matching initiating and implementing

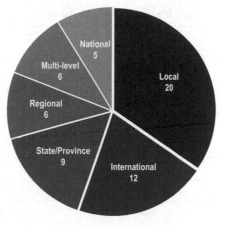

Figure 2.2 The Scope of Experimental Activity

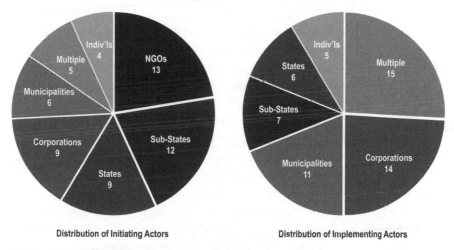

Figure 2.3 Distribution of Initiating and Implementing Actors

actors (30 of the 58 experiments are initiated and implemented by the same type of actor). For instance, the Transition Towns movement in the United Kingdom is an initiative designed by municipalities to be implemented by other municipalities. Yet a not insignificant number (28 out of 58) of experiments are attempts by one type of actor to govern other types (or multiple to govern multiple) like the European Covenant of Mayors, where nation-states in the European Union initiated a program designed for European cities to implement. (See table 2.1 for the breakdown of which types of initiators are setting rules up for which types of implementers.) Most governments (regardless of level) do not join initiatives that ask them to implement rules that they did not devise, an aversion that grows with

Table 2.1   **Initiating and Implementing Actors**

	Implementing Actor							
Initiating Actor	States	Sub-Nat'l	Cities	NGOs	Corporations	Individuals	Multiple	**Total**
States	6	0	1	0	0	0	2	**9**
Sub-National	0	6	0	0	3	0	3	**12**
Cities	0	0	6	0	0	0	0	**6**
NGOs	0	1	2	0	3	0	7	**13**
Corporations	0	0	0	0	8	1	0	**9**
Individuals	0	0	0	0	0	4	0	**4**
Multiple	0	0	2	0	0	0	3	**5**
Total	**6**	**7**	**11**	**0**	**14**	**5**	**15**	**58**

the size of the government. Nation-states and major subnational units (e.g., U.S. states and Canadian provinces) only implement experiments they initiated themselves. Cities are more amenable to external facilitation: 5 of the 11 experiments implemented by cities were initiated by other actors. Corporations are also often the target of experimentation: 6 of the 14 corporate-implemented experiments were initiated by noncorporate actors and the rest are a form of self-regulation.

This brief look at the demographics of experimentation gives a sense of two key characteristics of the experimental world—its recent emergence and its widespread reach both geographically and in terms of political scales. But in order to more fully understand the contours of the experimental world and the kind of global response to climate change that is emerging within it, it is necessary to ascertain more than which actors are participating and when/where they decided to experiment and look to what the experiments ask participating actors to do.

What Do Experiments Do (Or at Least Claim to Do)?

To characterize what experiments do, vision statements, charters, annual reports, and self-reported activities (press releases and reports) were consulted for each experiment in order to characterize their general objectives and the specific actions they undertake, information that was confirmed through interviews where possible for a number of experiments. Using largely self-characterizations to gather information on experiments' activities entails some potential limitations that should be noted up front. Initiators of experiments may be prone to exaggerate their actions—the websites and press releases are, after all, for the sake of publicity. It is difficult at times to verify, from documentation available on websites, whether initiatives actually do what they say they do. It is difficult to observe how much energy and resources various participants (initiators or implementers) put into any particular experiment. However, the self-characterization of climate governance experiments does have its value, especially for the kind of broad analysis developed in this chapter. In fact, it is precisely the measures that experiments "advertise" that are of interest—I want to know what different groups consider to be the appropriate means and modes of responding to climate change and how they justify those choices.[12] This analysis is as much an investigation into the potential to be found in a global response to climate change built on experimentation as it is an evaluation of the achievements of these recent initiatives. Even if they do not follow through entirely or effectively on their advertised activities, we can still learn about the ongoing evolution of climate governance by examining what various groups are proposing as ways to move forward on climate change and the distribution of activities found across the spectrum of experiments.[13] The Experiments were classified on the following dimensions:[14]

1. *Market or Regulatory Orientation:* Experiments were classified as being oriented toward market mechanisms, regulatory mechanisms, or a mix of the two in their response to climate change. Market measures include emissions trading, investment incentives, and other means to put a price on greenhouse gas emissions or to expand markets for climate-friendly technology. Regulatory measures include mandating specific levels of emissions reductions, reporting of emissions levels, and setting standards for reporting emissions and measuring emissions reductions. Some experiments were coded as mixed, as the rules they made included both kinds of activities.
2. *Mitigation or Adaptation Orientation:* Experiments were classified as being primarily focused on mitigation activities that seek to curtail greenhouse emissions or adaptation activities that concentrate on preparing actors for the effects of climate change. Some experiments include both kinds of activities and were coded as mixed.
3. *Commitment:* Experiments were classified as voluntary when *implementing* actors have a choice about participating in the experiment and carrying out its activities or mandatory when implementing actors are ordered to join the experiment.
4. *Specific Activities:* Each experiment claims to undertake certain actions—initiating actors devise activities that implementing actors are supposed to carry out. These tend to be idiosyncratic and tailored to individual experiments. However, looking at the universe of what experiments are asking implementing actors to do, the variation is finite. Certain activities show up in a number of experiments. Table 2.2 displays the actions undertaken or

proposed in the experiments along with the number of experiments that engage or ask implementers to engage in each activity (note that a single experiment can include more than one activity).[15]

5. *Core Functions:* The information on the *specific* activities—the things that implementing actors are supposed to do when participating in the experiment—was then used to classify each experiment as having one or more core functions: *general* modes of responding to climate change (see table 2.3 for the distribution of core functions across the experimental world):[16]

   a. *Networking. Linking* constituent members for the purposes of information diffusion through: education, information exchange, sharing best practices, direct negotiations, and regular meetings

   b. *Planning.* Designed to *prepare* actors for significant climate reductions, including: cataloguing emissions, targets, creating action plans, undertaking risk assessments, producing certification standards, or setting funding criteria

   c. *Direct Action.* Efforts to *directly reduce* emissions among constituent members via: efficiency measures, offset programs, production chain improvements, rationing, mandated reductions, emissions trading, and technological development

   d. *Oversight.* Means of gauging members' actions, including *monitoring* provisions and *enforcement* mechanisms

Tables 2.2 and 2.3 provide a sense of how activities and functions are distributed over this population of experiments and demonstrate the great variety of activities being pursued in the experimental world. Yet reporting only the gross number of experiments that undertake specific activities or have certain core functions understates the extent to which the activity in the experimental world is truly experimental. We might expect that different kinds of actors would gravitate to different kinds of activities and core functions, for instance, corporate innovations would engage with certain actions and municipal initiatives others. Experimentation might be patterned by actor type. This was, in fact, my first assumption, and in the initial conception of this book, the chapters were going to be organized around actor types (i.e., a chapter each on city initiatives, subnational initiatives, corporate initiatives, and multilateral initiatives). However, it turns out that there is *no* correlation between activities or core functions and types of initiating or implementing actors. This can be observed visually in figures 2.4 and 2.5, which display the specific activities (2.4) and core functions (2.5) pursued by experiments initiated by different types of actors. The lack of relationship between actor type and activities/core functions was confirmed statistically with a chi square analysis that demonstrated the independence of these variables. The identity of the initiating and implementing actors provides no prediction or understanding of the kind of activities likely to be employed by

Table 2.2 **Experimental Activities**

Activity	Number of Experiments
1. Catalogue Emissions/Undertake Inventory	20
2. Set Targets/Formulate Action Plan/Do Risk Assessment	32
3. Efficiency Measures or Offsetting	15
4. Education/Information and Best Practice Exchange/Regular Meetings	49
5. Set Certification Standards/Funding Criteria	4
6. Mandate emissions reductions	7
7. Emissions Trading	8
8. Monitoring (of implementing actors)	16
9. Enforcement	7
10. Technology Development	7

Table 2.3 **Experimental Core Functions***

Core Functions	Number of Experiments
Planning (Activity 1, 2, 5)	40
Networking (Activity 4)	50
Direct Action (Activity 3, 6, 7, 10)	26
Oversight (Activity 8, 9)	16

*Refer to the Appendix for the coding of each experiment for the core functions.

an experiment. Thus, not only do we see experimentation in the macro sense—many initiatives with lots of possible rules—but we also see it in the micro sense: similar actors trying out different strategies.[17]

To this point, climate governance experimentation has been revealed as a recent phenomenon that exhibits significant diversity in the type of actors experimenting and the type of activities undertaken. There is clearly a lot being done by a lot of different actors. This descriptive analysis is revealing, but potentially troubling. If all is cacophony, if the collectively, the experiments are just a series of disorganized, random attempts at doing something to address climate change, then there is little reason to suspect that they will have a broader *collective* impact on the global response to climate change no matter how innovative or important any particular experiment might be. If there is no coherence to the experimental world, then the strategy for studying these initiatives should be to examine them individually to see which individual experiments might have a transformative

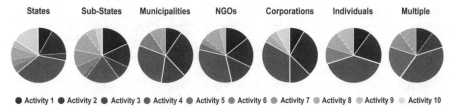

Figure 2.4   Activities (Refers to Table 2.2) Employed by Experiments Across Initiating Actors

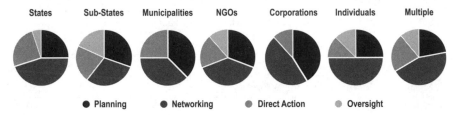

Figure 2.5   Core Functions of Experiments Across Initiating Actors

impact. Yet, further examination of what experiments do reveals that there is some method to the experimental madness and that there is reason to suspect that experiments can have collective relevance for how climate change is governed. It turns out that the experimental world is significantly patterned and organized, even in the absence of conscious planning to make it that way.

The Common Foundation of Liberal Environmentalism

The first evidence of a patterned experimental world is a set of common characteristics that can be found across this population of climate governance experiments—a shared baseline that provides the boundaries for devising ways of addressing climate change outside the megamultilateral process. Climate governance experimentation entails a voluntary, market-oriented approach that is focused on mitigation rather than adaptation. This is the common foundation shared by the diverse experiments in the database.

Fully 54 of the 58 experiments are classified as voluntary initiatives. This classification reflects whether or not implementing actors have a choice about participating in the experiment—all 58 are voluntary in the sense of arising through the voluntary association of initiating actors. This is not a surprising result, given the definition and nature of climate governance experimentation. Climate governance experiments emerge in the interstices of traditional polities and almost by definition lack the authority to command participation. C40 has no means of forcing the world's largest cities to participate. No one can be forced to join a CRAG. When actors initiate an experiment, they are inherently acting in a voluntary manner—the notion of being forced to experiment is nonsensical. Experimenters work to attract implementing actors, but for the most part they must do so through means other than coercion and command. This is the essence of what Hajer has called making policy without a polity[18] rules are being made and activities undertaken in the absence of legal authority to do so;[19] participation in governance experiments tends to be voluntary.

The four exceptions to the voluntary rule serve to reinforce its validity—they are the emissions trading venues initiated by subnational governments (U.S. states and Canadian provinces): the WCI, the Regional Greenhouse Gas Initiative (RGGI), the Midwestern Greenhouse Gas Reduction Accord (proposed, not yet officially agreed to), and the Ontario-Quebec cap and trade venue (ultimately abandoned as these provinces joined the WCI). These experiments buck the voluntary trend, in a way, because while the initiating actors (U.S. states and Canadian provinces) came together voluntarily to form these initiatives, they have the political authority to make participation for *implementing* actors (mostly corporations and in the case of RGGI electrical power generators) mandatory. The U.S. states and Canadian provinces do not have authority over each other, but they do have authority over the corporations and citizens within their borders. These traditional polities are making rules in a novel way—there is no set institution within which U.S. states and Canadian provinces collaborate—but they are able to use their traditionally sourced authority to make implementation *within each state/province* mandatory. While substate governments initiated these experiments voluntarily, implementing actors do not have a choice. The other experiments in the database either lack this coercive authority or, as in the case of alternative multilateral initiatives like the Asia Pacific Partnership or Methane to Markets program, choose not to exercise it when designing and implementing climate initiatives.[20]

Another way to consider the voluntary nature of governance experiments is to look at oversight and enforcement mechanisms. It may be voluntary to join an experiment, but perhaps once in, implementing actors are required to abide by the rules. While 16 of the 58 experiments have provisions for some kind of monitoring of participants, only 8—the 4 emissions trading venues mentioned above along with CRAGs, the Chicago Climate Exchange (CCX), Climate Savers, and the European Covenant of Mayors—have explicit enforcement rules with some penalties for noncompliance.

Yet the voluntary nature of experimentation may not be as clear-cut as it first appears. While virtually all experiments are voluntary, some experimental designs look to harness the political authority of various governmental actors to implement voluntary measures in an authoritative manner. The political authority of subnational actors makes them an attractive venue for the experimental designs of other actors, like NGOs. Experiments like ICLEI's CCP and The Climate Group's cities, states, and regions programs look to network and facilitate sub-national action precisely because they have the political authority to take substantial action on climate change. The Clean Air Partnership—the Toronto NGO that created the Alliance for Resilient Cities and the North-South Climate Change Network—uses the authority it has as a provider of expertise and creator of a network to attempt to influence how municipal authority is used in the province of Ontario. The Clean Air Partnership works to link civil servants and midlevel managers in its two networks and inform them about best adaptation practices as a way to catalyze change in cities and the province.[21] The Climate Group fully recognizes how cities and states can serve as useful laboratories for climate technology and that these entities have a unique ability to scale up technological deployment.[22] This is a relatively well-understood dynamic of environmental politics in the U.S. federal system. Rabe has chronicled the role of state bureaucrats in the evolution of U.S. climate policy.[23] Further, those involved in designing state-level emissions trading venues at the World Resources Institute recognize that environmental policy in the United States often grows from the state to national level.[24]

In fact, being voluntary does not mean that implementing actors will not comply with the rules of experiments or will fail to act out the experiments' core functions. Traditionally, environmental policy relies on the coercive authority of the government to ensure compliance. But compliance with experimental rules can arise from a range of sources, and many of these experiments represent the coming together of like-minded actors in ways that require less coercion.[25] Rosenau notes that authority is found in the "readiness to comply with directives . . . on a legitimacy that has been accorded the authorities by those toward whom their compliance efforts are directed."[26] Translated to climate governance experiments, whether or not the initiatives are voluntary, initiating actors may be authoritative, even if they lack the coercive legal authority inherent to governments. Whether because of expertise found in experiment initiators or because implementing actors share a common goal with initiating actors, or because of incentives that experiments provide to comply, the voluntary nature of experimentation does not necessarily preclude compliance.

In addition, while strict oversight and enforcement is the exception rather than the rule, public reporting and accountability are becoming part of the governance experiment world. Many experiments promote public reporting of implementation actions, and monitoring/accountability are of concern among the actors initiating experiments. The American College and University President's Climate Change program (ACUPCC) has a means for delisting colleges and universities that fail to adhere to their commitments and standards.[27] The Carbon Disclosure Project's entire raison d'être is based around public reporting—they urge large corporations to publicly disclose their carbon footprints as a way to inform investors about the climate exposure. The Partnership for Climate Protection (Canada's version of the CCP) is working to convince member cities that monitoring and reporting their climate activities is good for them. Further accountability within experiments is increasingly demanded by those that fund climate initiatives.[28]

The second common characteristic across the governance experiments is market orientation. Market mechanisms—pricing carbon, emissions trading, economic incentives, and the joint pursuit of greenhouse gas emission reductions and profit—dominate the ways experiments answer the question of how climate change should be appropriately addressed. Most experiments engage "carrot approaches" that seek to demonstrate how there can be economic co-benefits to taking action on climate change (e.g., The Climate Group has repeated a "Carbon Down, Market's Up" mantra since its inception)[29] like efficiency, stable carbon pricing, and/or technological development. Of the 45 experiments whose orientation could be ascertained, 40 have a market orientation or call for a mix of regulatory measures along with market measures. The Investor Network on Climate Risk maintains a position that is fairly common in the experimental world when it claims that "while climate change is clearly a pressing environmental—and perhaps even moral—issue, it is also inherently a business one."[30]

Only 5 experiments were coded as solely regulatory in nature, and again, these exceptions prove the market-orientation rule. Four of these five are carbon registries (American Carbon Registry, California Climate Action Registry, Greenhouse Gas Registry, the Climate Registry) that regulate and mandate the disclosure of information about carbon inventories and credits from their participants. But of course, the information being registered is most often designed to be used in subsequent market-oriented activities like trading, investment decisions and insurance calculations. The American Carbon Registry, for instance, is technically classified as regulatory because it sets standards that regulate what kind of emissions reductions can be registered, deciding what counts as a ton of greenhouse gas emissions reduced and accounting for and verifying the reductions reported by its implementing actors. Yet the registry has a wholly market-oriented purpose—its motivation is to provide part of the infrastructure for a carbon market. The world of climate governance experimentation is a market-oriented world.

This convergence around the ideas of voluntary association and market orientation is a classic example of what Steven Bernstein has called the compromise of liberal environmentalism—a compromise that "predicates environmental protection on the promotion and maintenance of a liberal economic order."[31] Climate governance experiments are, for the most part, advocating climate action by making the case that such action is or will be economically beneficial. Collectively, experiments are attempting to shift the focus of the discussion about the response to climate change from the costs of reducing emissions to the (economic) benefits of reducing emissions. The Business Council on Climate Change, a relatively small experiment initiated in the San Francisco Bay Area, is explicit about this commitment when they claim to be providing "a model of commercial climate stewardship" and building "a business movement for climate stewardship."[32]

In this sense, then, experimentation is *not* a revolutionary challenge to established governance mechanisms. Observers have noted the underlying market/ liberal orientation of multilateral environmental governance found in discussions of sustainable development, public-private partnerships, and the climate negotiations themselves.[33] While there are certainly calls for radical rethinking of nature-economy-society relationships, by and large these calls do not substantially influence the experimental governance world. Instead, climate governance experiments tend to reinforce the notion that the transition to a carbon neutral world should and will take place through market-oriented means rather than through a radical rethinking of social and economic structures. On the one hand, this may allow experiments to be taken more seriously by politicians, industry, and the broader society, given their economic orthodoxy. They tend to ask for less 'sacrifice' and focus instead on the benefits of moving to a low carbon world. On the other hand, as the protesters at the 2009 UNFCCC COP in Copenhagen who decried the advance of green capitalism would no doubt point out, it is not entirely clear how effective the incremental changes that arise from market mechanisms can be, or what impact such mechanisms will have on equity concerns that drive important debates over the global response to climate change.

Finally, mitigation of climate change is the overwhelming focus of experimentation. Of the 58 experiments, 40 focus exclusively on mitigation, while another 16 pursue a mix of mitigation and adaptation efforts. This result could emerge from possible selection bias in the definition of climate governance experiments. Where adaptation efforts exist, they are often not couched explicitly in climate change terms or they deal exclusively with local concerns and thus not cross political boundaries. They fall into the fuzzy area of sustainable development or are more focused on specific problems like flood control, soil erosion, land management, emergency response, even health.[34] In addition, adaptation has only recently emerged as a key feature of climate governance (in the multilateral level or beyond). While many actors in the international system, especially small island states that face an obvious existential threat from climate change, have consistently advocated for more attention on adaptation, only in the last few years has the spotlight been shone on this aspect of responding to climate change—especially among city-led and -implemented initiatives and experiments coming from insurance and investor groups.[35]

While the experiments captured in this database are characterized by a focus on mitigation, adaptation will be an increasingly important component of the global response to climate change in the years to come. The world is already committed to some climate change, given the concentrations of greenhouse gases already in the atmosphere and that some signs of climate change, like the melting of arctic sea ice, are already evident. We are already adapting to climate change. In addition, while the experiments' efforts concentrate on mitigation and reduction of greenhouse gas emissions, at least some were initiated because the effects of climate change already being felt. Mary MacDonald from the city of Toronto noted that one motivation for city-led experimentation was that mayors, as local level politicians, can see the effects of climate change most readily.[36] Further, the 2009 Copenhagen negotiations enshrined adaptation as a multilateral concern. One of the few bright spots in what emerged from these meetings was the $100 billion pledge to help Southern countries deal with effects of climate change.[37]

Governance Models in the Experimental World

The master narrative for climate governance experimentation thus comes into focus. Climate governance experimentation is a liberal environmental phenomenon that emphasizes voluntary, market-oriented mitigation as the global response to climate change. This common foundation is the first element of organization in the experimental world. Further analysis of the core functions that constitute these experiments reveals additional structure. The diverse activities and core functions evident in tables 2.2 and 2.3 are not an expression of randomness. Instead, distinct groups of experiments, or what I call governance models, are identifiable by their commitments to specific individual core functions or combinations of core functions (see Box 2.1 for information on the cluster analysis methods used to identify groups).

The first governance model, *networkers*, includes experiments whose only core function is to network. The center of gravity of the second model is planning. Experiments following this model, *infrastructure builders*, either just plan, or plan and network as their core functions. The third model is distinguished by a focus on taking direct action. Experiments following this model, the *voluntary actors*, all have direct action as a core function while they also network and plan (at least some of them). Finally, the experiments in the fourth model, *accountable actors*, employ all four core functions—networking, planning, direct action, and oversight.

The world of climate governance experiments thus is far from a random assortment of initiatives and is actually quite structured, with a common philosophy underlying all the experiments, and clear functional differentiation. Instead of thinking of experiments as idiosyncratic, we can think of them as playing specific roles or embodying specific identities in the global response to climate change: networkers, infrastructure builders, voluntary actors, and accountable actors. A brief discussion of the characteristics and constituents of each governance model is warranted before moving to further analysis of the structure and impact of the experimental world in the subsequent chapters.

Box 2.1 **Cluster Analysis**

The groups of experiments were identified through cluster analysis, a statistical procedure for inductively revealing patterns in a set of data that is appropriate when the goal is to "identify a set of groups which both minimize within-group variation and maximize between-group variation."[38] Essentially, this eponymous technique takes as input a series of values on different variables for each case in a data set and returns natural groups of cases that fit better together than they do with other groups of cases.[39]

The challenge is choosing variables to define the clusters, as there are few a priori means of doing so. Given the results of the chi square analysis discussed above, I suspected that including both actor variables and activities would not produce stable clusters precisely because the descriptive analysis showed these variables to be independent of one another. This was confirmed in early cluster analyses. Including all 10 specific activities in the cluster analysis was untenable because too many defining variables rendered it impossible to define stable clusters.[40] Using the 4 core functions (networking, planning, direct action, and oversight) did produce stable clusters (table 2.4).[41] These 6 clusters themselves group naturally into 4 metaclusters—the governance models mentioned above—based on which core function is the distinguishing feature.[42] The shading in table 2.4 indicates the four governance models.

Table 2.4 **Core Functions in Clusters**

Cluster	Number of Experiments	Networking	Planning	Direct Action	Oversight
1	11	X			
2	7			X	(4/7)
3	14	X	X		
4	7	X		X	
5	11	X	X	X	
6	8	X	X	X	X

Note: An "X" denotes that all experiments in the cluster encompass noted activities.

Model 1: The Networkers

> The Climate Group is looking beyond Kyoto, which is a first step in driving reductions. . . . We know that there are many leading companies and governments dedicated to meeting or exceeding those targets. By bringing the key players together for the first time, we believe that the world can turn the corner on climate change.
>
> —Steve Howard, CEO of The Climate Group (April 1, 2004)

Experiments classified as networkers restrict their activities to one core function and respond to climate change by fostering information exchange among participants, undertaking educational activities, meeting to discuss best practices, or even just putting actors interested in climate change together. This does not imply that the actors that initiate networking experiments are lax on climate change. Quite the contrary, most initiators of networking experiments expect, or at least hope, that implementing actors will move beyond exchange of information and toward significant emission reducing actions. The logic of this governance model is well expressed by Steve Howard in the epigraph to this section.[43] Fundamentally, networking experiments consider that the problem with the response to climate change is not lack of will but poorly distributed information and resources to take action. The solution is to bring actors together—physically and/or virtually—to spread ideas and practices that can be implemented. The goal of networking is to catalyze action outside the jurisdiction of the experiment itself through peer-to-peer motivation and learning.

Networking experiments range from the very individual-scale Edenbee, an online social networking platform for climate action with around 4,500 members, to the very local Ontario-based Alliance for Resilient Cities, to regional programs like the Network of Regional Governments for Sustainable Development, and to major NGO-corporate alliances like The Climate Group. A major multilateral initiative is even included in this category—the Asia-Pacific Partnership. A range of different actors have decided that an appropriate way to address climate change is to link like-minded actors—to network.

The concept of networking has become popular, and its ascendance spans academic writing and the popular press.[44] Networks are alternatively conceived as a new principle of sociopolitical organization (in contrast to both hierarchies and markets), as a new tool of social analysis (social network analysis, which began in sociology, has expanded rapidly in the social sciences), and as a new metaphor for the logic of the modern epoch (the networked society metaphor has emerged from the revolution in digital and communications technology).[45] Networking has become a familiar part of the modern lexicon, so the fact that a large majority of the initiatives in the climate experiments database (50 out of 58) are engaging in networking activities is no surprise. But the focus here is on those 11 that were coded as *exclusively* engaging in networking. The kind of activities they engage in is instructive for thinking about the role of networking in the broader global response to climate change. Box 2 gives a brief taste of what the networkers are doing.[46]

These are diverse activities engaging a wide range of actors, but two essential networking activities can be distilled from these activities. The first is education, broadly conceived. Most of the networking experiments have been designed to disseminate information and ideas about climate change. Whether this is couched in terms of sharing best practices, learning about solutions, or acquiring up-to-date scientific/policy developments, networking experiments serve to diffuse knowledge about climate change and ways to address it. A testimonial on the 2degrees website captures the essence of the networking experiment logic:

> 2degrees provides a quick and effective link to numerous organisations and individuals who want to share information, ideas and experiences. It's these sorts of links that are invaluable when you're trying to initiate change within an organisation, especially when it comes to reducing our dependence on carbon.[47]

The second is community building. Networking experiments work to establish relationships between actors who otherwise might not be in contact. Most networkers also go beyond merely establishing relationships to active community building initiatives that include mutual support activities. This can be as simple as Edenbee's members sharing their experiences with each other to The Climate Group's more strategic deployment of its corporate members in support of subnational governments' actions.[48]

Networkers are attempting to build communities of like-minded actors. The linkages are not merely functionally oriented; they are instead at least partially made in the service of growing communities of actors across established borders that are committed to progressive action on climate change. In this sense, networking is a means to an end. Education and community building efforts are established to move the educated communities in particular directions. Most networks are very explicit about this. On its Web page, The Climate Group trumpets, "over the next five years, The Climate Group's goal is to help government and business set the world economy on the path to a low-carbon, prosperous future. To reach this goal, we've created a coalition of governments and the world's most influential businesses all committed to tackling climate change."[49] The International Climate Action Partnership is explicitly building a community that it hopes will build a global carbon market. Other experiments have less explicit goals. For instance, 2degrees certainly has a broad sustainability goal in

---

Box 2 **The Networkers**

2degrees: This initiative sets up infrastructure for "sustainability professionals" to connect. They have a number of climate related networks. Members of the network can search for content posted by others, connect with specific individuals, ask experts questions, watch or host webinars. A majority (2/3) of members are corporate, but members also from NGOs, academia, and government (1/3). Multiple sectors are represented.

Alliance for Resilient Cities and the North South Climate Change Network: Both of these initiatives were created by the Toronto-based NGO the Clean Air Partnership. They have very similar mandates to 2degrees (webinars, information sharing) but target local officials in Ontario and network both academics and civil servants.

Asia Pacific Partnership for Clean Development and Climate Change: This is really a set of sectoral networks developed by nation-states. The focus is on information exchange between various sectors, and they have set up a number of working groups (cement, aluminum, coal, steel, buildings, energy). The role of the experiment is to facilitate information exchange and provide an environment for finding common solutions.

Edenbee: This is essentially Facebook for carbon cutters (over 4,500 members as of January 2010). It is a decentralized, mutual support group for people who have similar goals. There are goals listed, but they emerge from the members reporting what they do. The raison d'être of the site is to build a community of like-minded individuals.

Evangelical Climate Initiative: This is an education focused experiment and is designed to inform evangelical leaders and church members about climate action and to spread individual responses.

International Climate Action Partnership: This is another education-focused experiment where members participate in workshops, seminars, and joint research on the design and functioning of carbon markets, including the assessment of linking possibilities between such existing markets so as to contribute to the creation of a global carbon market.

Network of Regional Governments for Sustainable Development: This is designed to be a voice for and to represent regional governments at the global level in promoting sustainable development by exchanging information and experience. It also seeks to promote collaboration and partnerships at the regional level.

Renewable Energy and Energy Efficiency Partnership: This is an interesting experiment that undertakes some project functions on its own and organizes networking functions for other actors. Their networks are designed to deliver expert information and knowledge of best practices.

The Climate Group: This experiment brings together subnational governments and corporations to spread best practices and solutions. They have a number of mutual support functions as well that are designed to enhance the potential of network members to catalyze market and political transformation. Of late they have begun implementing pilot technology deployment projects and may evolve into the voluntary actor model.

World Business Council on Sustainable Development: This is a CEO-led initiative that has dual roles advocating for business positions and serving as a means for corporations to share knowledge of best practices on sustainable development and climate change.

mind but is less directional. A number of networks of experts and "sustainability professionals" are available through 2degrees, and members can propose new networks depending on their needs and goals.[50] Edenbee is perhaps the most open—the goals for the network emerge organically from goals that individual members have set for themselves. Other than having a generic goal of addressing climate change, Edenbee serves as an archive of climate goals evident across the community, rather than guiding the community it has established in a specific direction. What networking experiments do—education and community building—is not equivalent to what they hope to accomplish.

Model 2: The Infrastructure Builders

The first step towards managing carbon emissions is to measure them because in business what gets measured gets managed. The Carbon Disclosure Project has played a crucial role in encouraging companies to take the first steps in that measurement and management path.

—Lord Adair Turner, chair, UK Financial Services Authority

Responding to climate change by reducing emissions of greenhouse gases and ultimately weaning the global economy from its reliance on fossil fuel for the production of energy requires an enormous amount of preparation. We need to be able to accurately measure and track carbon emissions from specific entities (whether cars or factories or something more abstract like cities and corporations).[51] We have to decide where to count emissions, for instance deciding who is responsible for the carbon emissions that go into the production of electricity, the generators or consumers. We have to understand the various ways climate change inflicts risk on business, governmental, and nongovernmental actors. We need to understand what types of activities can produce efficient reductions and how to implement them. If there is to be a carbon market, we have to do a number of these activities while also accounting for amounts and prices of avoided emissions—commonly known as offsets.[52] More simply, we need infrastructure, in terms of ideas, institutions, and technologies to *facilitate* the global response to climate change.

A number of climate governance experiments have taken on this challenge, deciding that their role is to build the capacity of diverse actors to engage in the global response to climate change. Of the 58 climate governance experiments, 40 have been classified as engaging in planning. The 21 included in this governance model are unique in that planning is their distinguishing feature. Either it is their only core function, or they undertake both networking and planning. The experiments in this model have pursued a strategy of preparing other actors for action on climate change or providing the means to facilitate action. They are laying the groundwork for a global response to climate change in tangible ways. Box 3 provides a synopsis of the activities of the infrastructure builders.[53]

Two broad categories of infrastructure building are evident in these brief descriptions—measuring greenhouse gas emissions/reductions and action planning. In the discussion of emissions reductions, numbers like a 20% reduction below 1990 levels by the year 2020 are thrown around quite casually—not that the commitment is casual, but it is too often a casual assumption that we know how to measure such reductions accurately. Ironically, it is relatively easy to calculate national-level emissions (nation-states can readily gather aggregate fuel use and electricity generation), but carrying out domestic regulations, building a carbon market (generating both producers and consumers of emissions permits), or participating in some more stringent experiments that mandate reductions requires measuring greenhouse gas emissions at a more fine-grained scale—at the level of factories, corporations, cities, households. Infrastructure building experiments have emerged that attempt to fulfill this need in a number of ways.

A number of registries and programs offer means and protocols for members to measure their greenhouse gas footprints. The Climate Registry, discussed in greater detail in chapter 4, is a key example of this function. "The Climate Registry establishes consistent, transparent standards throughout North America for

Box 3 **The Infrastructure Builders**

American Carbon Registry: This experiment sets standards for how to measure greenhouse gas reductions in the production of offsets for voluntary carbon markets (markets for buying and selling greenhouse gas reductions where no emissions reductions are mandated by governments—which would be regulated markets). They also provide registry services where offset producers and consumers can list their greenhouse gas "holdings."

Business Council on Climate Change: This experiment specializes in company assessments of carbon footprints, employee programs, and corporate best practice sharing networks.

California Climate Action Registry: Originally designed as an inventory for California businesses and governments to list their carbon footprints, in 2006 the California registry split into the Climate Action Reserve, which works on setting offset standards, and the Climate Registry, which took the inventory function nationwide in the United States.

Carbon Disclosure Project: Dedicated to getting companies to report their carbon footprints and exposure to climate change risks, this experiment gathers information intended to be used by institutional investors.

Carbon Sequestration Leadership Forum: This multilateral experiment focuses on policy planning and networking around carbon sequestration technology.

CarbonFix: Another standard-setting experiment that focuses on forest carbon offset projects.

Climate, Community, and Biodiversity Alliance: This unique standard setter for forest carbon offset projects sets standards for social and biodiversity impacts of climate offset projects.

Climate Neutral Network: This experiment is mostly a best practice sharing network like those in Model 1, but they do require members to have a climate neutral strategy in place to join the network.

Community Carbon Reduction Project: A flexible environmental management standard/system wherein participants are expected ultimately to contribute to a collective goal of reducing emissions 60% by 2025. Still, each participant (individual, community, or corporation) is free to decide its own actions and commitments as contributions to this goal.

Global Greenhouse Gas Register: This now defunct experiment was an effort to create a global emissions inventory program for corporations headquartered in countries that lacked Kyoto goals.

ICLEI CCP Campaign: A planning experiment for municipalities, the CCP set up benchmarks and a support system to aid cities in developing their climate change action plans and actions.

Institutional Investor Group on Climate Change; Investor Network on Climate Risk; Investors Group on Climate Change: All three of these investor-oriented experiments are focused on getting institutional investors variously in the United States, United Kingdom, and Australia to include climate risks in investment decisions.

Klimatkommunerna: Similar to ICLEI, in this association of Swedish cities and regions members develop inventories, targets, and action plans. They are unique in that they also have some oversight functions (in cluster 2, they are one of the 4 that do so), in that members are required to regularly report their progress.

Methane to Markets: This multilateral experiment is similar to the Carbon Sequestration Leadership Forum—an information sharing and policy development program for recovering and using methane as a fuel.

National Association of Counties County Climate Protection Program: This initiative among U.S. counties is modeled on various cities programs—members undertake greenhouse gas emission inventories and develop county-wide climate reduction action plans.

Southwest Climate Change Initiative: A rather vague experiment that is now defunct and superseded by the WCI; members pledged to develop greenhouse gas emission measurement and reporting programs, to promote climate change mitigation actions, and to explore energy efficient technologies and clean and renewable energy sources that enhance economic growth.

The Climate Registry: This is the major climate inventory initiative in North America, having developed protocols for reporting greenhouse gas footprints for subnational governments, corporations, and NGOs. It is run by a board composed of representatives from U.S. states and Canadian provinces.

Transition Towns: This is mostly a networking and information exchange program for local communities to prepare to transition to a low carbon future, but it includes development of action plans to prepare for transition to low carbon energy.

Union of the Baltic Cities Resolution on Climate Change: Similar to other city planning networks, Union cities pledge to plan climate actions—both mitigation efforts and adaptation contingencies.

businesses and governments to calculate, verify and publicly report their carbon footprints in a single, unified registry."[54] Another set of infrastructure builders also looks to record greenhouse gas emissions but in order to facilitate a greater understanding of climate change exposure. The Carbon Disclosure Project and the three investor-oriented experiments all seek to increase the information available about corporations' greenhouse gas profiles in order to leverage investor decisions toward climate-friendly actions. Finally, there is a set of standard setters, mostly for the offset markets. Initiators of these experiments observed that in the early stages of the offset market, almost anyone could claim to be reducing greenhouse gas emissions and offering to "sell indulgences" to those who themselves wouldn't or couldn't reduce.[55] Worried about the integrity of the carbon market, a number of programs and experiments have emerged (Climate Action Reserve, American Carbon Registry, Voluntary Carbon Standard, Carbon Fix, Climate, Community, and Biodiversity Alliance) to provide protocols and methods for measuring, verifying, and recording greenhouse gas reductions (discussed at greater length in chapter 6).

Action planning is the second type of infrastructure building that has emerged in the experimental world. The CCP is perhaps the most well known of the action plan–oriented experiments, with its benchmarks for municipal action on climate change. Megan Meaney, director of Canada's version of the CCP, observed that cities want to join the program because they want to participate in a standard—the CCP experiment provides the model for pursuing municipal climate action.[56] The provision of models and requiring the development of action plans by implementing members of experiments is a key aspect of about half of the infrastructure building experiments. When combined with the networking functions that many of these experiments also undertake, the infrastructure building activities have the potential for wide impact.

Similar to the networkers, infrastructure builders have larger goals and the activities required of implementing actors are instrumental to these. The infrastructure building experiments lay the foundation and provide the tools for a global response to climate change. These experiments facilitate climate action rather than requiring it. This can lead to charges of ineffectiveness—a charge that has hounded the CCP program[57]—and experiments have not always been able to track the results that come from all the planning and measuring of greenhouse gas emissions.[58] The worries are that too few are taking advantage of the tools infrastructure builders provide and that planning has yet to lead to significant action. It may be too early to tell, because the hope is that planning and providing the tools to measure and reduce greenhouse gas emissions will have a nonlinear effect—a threshold will be reached whereby emissions reductions can be reduced rapidly because the infrastructure for doing so is ready.

Model 3: The Voluntary Actors

While the first two models can be considered ways of building foundations for more tangible action on climate change, experiments in Model 3 directly, though voluntarily, act to reduce greenhouse gas emissions, in addition to planning and networking. This group includes major multilateral and bilateral initiatives among nation-states and subnational actors. These experiments call on participants to plan, network, and take action but without obvious means of accountability. As noted, while all experiments are essentially voluntary, the experiments in this model go beyond voluntary networking and planning to include direct actions aimed at reducing greenhouse gas emissions.

Voluntary action in the environmental realm has received a great deal of attention in the last few decades and a significant amount of criticism. This kind of action has often been related to the corporate social responsibility movement. Corporations have been at the forefront of advocating for voluntary responses to environmental problems—at least in part as an effort to avoid regulation.[59] But the experiments in this model are not only corporations. On the contrary, subnational actors, nation-states, and NGOs have also initiated experiments in this model. The characteristic that makes many of these experiments different from voluntary initiatives in other areas is that they have been founded by like-minded actors that require less in the way of enforcement to achieve compliance. Many of the voluntary action experiments are undertaking action not so as to avoid mandatory regulations but so as to go beyond existing regulations or as a way to press for more stringent regulation.

The three major city networks (the Climate Alliance, C40, and the U.S. Mayors' Conference Climate Protection Agreement) in this model, for instance, are all associations of actors that came together precisely because they were like-minded in their dedication to going beyond what was being done in other areas and as a way to showcase what could and should be done on climate change in broader political jurisdictions. For instance, C40 does not have formal oversight rules, but Mary MacDonald, a key advisor to former Toronto mayor David Miller (the former chair of C40) stresses that "mayors know mayors" and that "mayors make agreements with mayors and hold each other accountable in their own way."[60] Thus voluntary does not necessarily mean a lack of compliance or accountability, even if no formal measures are in place in the experiment.

Steve Schiller, a former board member of the California Climate Action Registry and the Climate Registry, is convinced that there is a continuum of progress on climate action and that voluntary efforts, like the registries he has worked on and the experiments categorized here as voluntary action are stepping stones toward mandatory climate policy—voluntary reporting gives way to voluntary action which leads to mandatory action with low targets and finally mandatory action with high targets.[61] Box 4 briefly describes the kinds of action being undertaken voluntarily in the experimental world.[62]

At first glance, these direct action activities appear to be all over the place—efficiency measures, infrastructure measures, public transit, building codes, transportation measures, information technology deployment—and of course this is on top of the networking and planning in which most of these experiments also engage. But that is part and parcel of experimentation. These experiments seem to be held together as a governance model more by what they do not do—accountability—than by what they do undertake—a wide variety of possible actions. Yet a second look reveals that while these may be the "do something" experiments, there is a common ethos at work here in terms of experiments defining the type of action by not just doing anything but tailoring action to fit the needs of implementing actors and doing something to set an example.

It makes perfect sense for an experimental model to emerge that is customized for its implementing actors—all experiments are to some extent, but it is perhaps even more notable among voluntary action experiments. In the absence of a top-down program of climate policy, what emerges is implementing actors pursuing activities that fit their more individualized needs. The experiments facilitate and support this and provide structure for doing it but are not holding actors accountable for what they agree to do. Either authority to monitor and enforce experimental mandates is lacking, or, enforcement measures have yet to be fully thought through—precisely because these experiments are bringing together like-minded actors. In addition, many, if not most, of these experiments do not see themselves as the locus for the global response to climate change. On the contrary, many are pursuing these voluntary actions so as to set an example (e.g., ACUPCC, Climate Wise, Conference of New England Governors, Cool Counties, U.S. Mayors' Conference Climate Protection Agreement) or to explicitly build momentum toward larger responses (e.g., various MOUs, Major Economies Forum).

Box 4 **Voluntary Action Experiments**

ACUPCC: Members of this group pledge to eliminate their campuses' greenhouse gas emissions through both short-term actions (e.g., implementing green building standards, energy efficient purchasing,

travel offsets, increase public transit access, renewable energy purchasing policy, waste reduction programs) and integrating sustainability into the curriculum.

Bilateral Partnerships and Memoranda of Understanding: A number of nation-states and U.S. states have negotiated specific partnerships
around climate change. Some of these had less than progressive
aims. For instance, Australia set up bilateral agreements with the
United States (February 2002) Japan (May 2002) New Zealand (July
2003) China (August 2004), and South Africa (2006) while it was
repudiating the Kyoto Protocol. Others are oriented toward catalyzing great action. The United Kingdom initiated a series of MOUs
with U.S. states in order to support climate action when the U.S. federal government was loath to take action. Similarly, the state of California negotiated MOUs with various nation-states in order to
further its climate work. Recently, great powers have engaged in this
activity as well, with the United States and China signing a bilateral
MOU on climate and energy. In addition to networking, these
partnerships are also designed to foster technological development
and deployment.

C40: This group of 40 of the largest cities in the world began with pledges
to enhance sustainable procurement at the municipal level and has
expanded into renewable energy and other technological deployment
programs in cities.

Carbon Finance Capacity Building Programme: This partnership of experiments and other actors (C40, the Clinton Climate Foundation, the
Intergovernmental Panel on Climate Change, the World Bank) works
to integrate cities into carbon markets, helping municipalities to develop emission reduction projects and turn them into assets for possible trading.

Climate Alliance of European Cities with Indigenous Rainforest Peoples: City
networks are well represented in this governance model, and this experiment is similar to C40 and the CCP program. It is centered in
Europe, and members pledge to reduce their emissions by 10% every
five years.

ClimateWise: Major insurance companies pledge to include climate risk
analysis into their activities, reduce their own emissions, and incorporate climate change concerns into investment strategies.

Clinton Climate Initiative: In addition to providing support to the C40
group and facilitating implementation of its programs, this experiment also works on technology demonstration projects.

Conference of New England Governors and Eastern Canadian Premiers Climate Change Action Plan: Canadian provinces and U.S. states have pledged to craft climate action plans, reduce public sector emissions by 25% from an established baseline by 2012, and reduce emissions from both the transportation and electricity generation sectors.

Connected Urban Development Programme: What began as a corporate social responsibility project for the Cisco corporation has turned into a large-scale pilot project involving a number of cities deploying information technology in ways that reduce municipal greenhouse gas emissions.

Cool Counties Climate Stabilization Initiative: Implementing counties agreed to reduce emissions from their own operations as a way to demonstrate leadership. They also pledged to become climate resilient.

e8 Network of Expertise for the Global Environment: Major electric companies pledge to implement energy efficiency measures.

EUROCITIES Declaration on Climate Change: These cities have agreed to go beyond action planning to reduce municipal operation emissions, implement building retrofits, develop public transportation, and develop adaptation plans.

Major Economies Forum on Energy and Climate: This experiment may be the single biggest multilateral challenger to the Kyoto process. It began as a U.S. initiative to wrest control of climate policy-making away from the UN process. It has become the most important multilateral experiment for responding to climate change.

U.S. Mayors' Conference Climate Protection Agreement: In repudiation of the U.S. withdrawal from the Kyoto Protocol, hundreds of U.S. cities agreed to measure their emissions and reduce greenhouse gas emissions from municipal operations.

West Coast Governors' Global Warming Initiative: These U.S. states plan to engage in bulk buying of hybrid vehicles, increase the sale of renewable energy, adopt energy efficiency standards for products, and update state building codes for energy efficiency.

Model 4: The Accountable Actors

The climate governance model with the fewest experiments is the accountable actors, which pursue initiatives that engage all four core functions—networking, planning, direct action, and oversight. This is the least populated model because it asks, or rather demands, that participants do the most. What sets this model apart is not just the commitment to take direct action but also the existence of oversight measures that make that action accountable. The bulk of the 8 experiments in this model were initiated by subnational or supranational government bodies that have the authority to mandate or at least monitor participation in climate governance initiatives within their jurisdictions. The other 3, Climate Savers, CRAGs, and the CCX, are fascinating cases in which implementing actors voluntarily sign up for climate governance experiments that contain real provisions for monitoring and enforcing the rules.

Enforcement has always been a crucial issue in global governance. One of the key critiques of international law writ large is that it is very difficult to enforce international agreements and legal mandates. A case in point is the failure of Canada to live up to its Kyoto commitments (or even make an effort toward doing so). While environmental NGOs heap scorn on the Canadian government and international lawyers discuss the implications of being out of compliance, in practice there has been no consequence to Canada for its noncompliance. One of the biggest debates in the 2009 Copenhagen negotiations was over this very concern—monitoring and verification of pledges. The increasing salience of this issue was heartening to see, but the results of negotiations were not as positive. The Copenhagen Accord does not enshrine a strict policy of monitoring and verification of pledges and the only measure that the group that negotiated it could agree to was the each nation-state would monitor and verify its *own* commitments. Yet even in the experimental world which is largely based on voluntary cooperation, we see the possibility for legally enforceable pledges and emission reduction activities. The 8 experiments listed in box 5 pledge not only to take emission reduction actions but also to enforce the commitments of implementing actors.[63]

Perhaps the most striking feature of this model is the focus on emissions trading as the substance of the response to climate change. Of the 8 experiments, 5 are explicit emissions trading systems, while CRAGs employ a traditional emissions cap and penalty system and the implementers in the Covenant of Mayors and Climate Savers mainly pursue energy efficiency measures. In fact, going beyond strictly defined climate governance experiments, there has been a proliferation of interest in emissions trading as a central policy for addressing global climate change.[64] While the UN negotiations at Copenhagen in December 2009 floundered, the parallel discussions on the development and future of carbon markets and emissions trading demonstrated that the momentum behind this means of addressing climate change is still growing and that emissions trading may be poised to become a central piece of the global response to climate change (though developments in 2010 were disappointing for cap and trade enthusiasts). By its very nature, cap and trade initiatives must have enforcement—without it, the integrity of the market created is seriously compromised. The

Box 5 **Accountable Action Experiments**

CRAGs: Individuals in the United Kingdom, Canada, and U.S. communities self-organize to impose a collective cap on their group's emissions and individual rations that must be met enforced by fines paid to the group.

CCX: This is the first private emissions trading venue. Corporations and other actors that join the CCX agree to a modest reduction—by 2010, 6% below a baseline calculated using average annual emissions of 1998–2001. Actors can trade allowances or obtain offsets to meet their goals. Joining is voluntary, but the commitment is a legally binding contract.

Climate Savers: Corporations who join this experiment, initiated by the World Wildlife Fund, commit to energy efficiency measures or fuel-switching measures in an agreement that is negotiated with the World Wildlife Fund and monitored/verified by an independent third party.

Covenant of Mayors: These European cities pledge to implement a sustainable energy action plan in service of moving beyond the European Union's reduction goals. Failure to report on the plan or make progress toward the reduction objectives is grounds for being removed from the program.

Midwestern Greenhouse Gas Reduction Accord, Ontario-Quebec Provincial Cap-and-Trade Initiative, RGGI, WCI: These four experiments are the North American subnational emissions trading systems that have been discussed and/or implemented in the last decade. The RGGI is currently operating in the Northeast United States. The WCI is set to come online in 2012, engaging western U.S. states and a number of Canadian provinces. The Midwestern Greenhouse Gas Reduction Accord is nearing the end of the design phase. The Ontario-Quebec initiative was scrapped in favor of both provinces joining the WCI. All four of these are cap and trade systems whereby subnational governments decide on allowances for regulated entities and manage the trading system—including enforcement when regulated entities fail to meet achieve their reduction mandates.

emissions permits that are the substance of the trading system only have value is if there is a penalty for failing to stay within one's allowance. If a company can be a part of a cap and trade system but pay no penalty for going over its cap, then the worth of the tradable permits falls quickly to 0.

The two non-cap-and-trade experiments also demonstrate that other kinds of action can potentially be made accountable even in the absence of governmental authority. Given the range of experiments and activities encompassed in the previous voluntary action model, this is an interesting lesson to consider. Perhaps the progression that Schiller identifies, moving from voluntary to mandatory action is possible among experimental initiatives and enforcement may not be restricted to the governmental realm.

Conclusion

This chapter began the process of exploring the emergence and implications of experimentation by describing its contours. Experiments are not all the same (i.e., there is significant variation in which types of actors are experimenting as well as what rules and activities they devise and implement), but they are not simply random or idiosyncratic initiatives either. On the contrary, climate governance experiments are patterned. They hold in common a liberal environmental philosophy that seeks to reconcile economic growth with environmental protection, in that they focus on voluntary and/or market-oriented activities. There is also functional differentiation among the experimental initiatives. Looking at the climate governance experiments *collectively* reveals how the experimental world is organized.

What remains to be seen is whether this patterning has any relevance for the global response to climate change. The question is whether patterned governance experiments can be thought of as a system of governance—a decentralized means of responding to climate change—analogous to, but functionally different from, the multilateral system of governance. In order to address this question, we need to make sense of experimentation at two levels. First we need to understand individual experiments. Why are specific initiating actors experimenting and how do they choose to experiment in the manner they do? Some actors have come to see themselves not just as contributors to the multilateral response to climate change, but rather as developers of their own responses to climate change. Understanding experimentation requires explaining this choice.

The second is at the level of the collection of experiments. Of what is patterned climate governance experimentation an instance?[65] Is the patterning of experiments evident in this chapter merely a coincidence or even the result of a very human proclivity to see patterns even in randomness? Or is it indicative of an emerging organized, if decentralized, means of governance that does not rely on a top-down, multilateral treaty to catalyze action on climate change? Such a system would have unbelievable contrasts with the megamultilateral process. It would involve multiple actors and diverse rule-making practices, as opposed to set actors and an established, singular means of making rules. It would be flexible, whereas the megamultilateral process is tied to consensual decision-making. It would include a myriad of actors at the outset instead of at the tail end of implementing a treaty. It would have areas of questionable political authority instead of the standard authority of international law and sovereignty, but in bringing together like-minded actors around a range of activities, enforcement might be less of a significant issue. We now have a better sense of *what* is happening with climate governance experiments, but we need to pursue an (analytic) story that can account for the emergence and patterning of climate governance experiments as well as provide us with guidance for what to expect from climate governance experiments.

3

Making Sense of Experimentation

Regional initiatives are incredibly important.
—Nancy McFadden, Senior Vice-President
—Pacific Gas and Electric UNFCCC COP
meeting at Bali (December 2007)

We [cities] have a sense of urgency that seems to be
lacking [among national governments].
—Micki Gavron, deputy mayor of London,
at UNFCCC COP (December 2007)

Getting involved in a crag is a way of pioneering—
experimenting with how to make low carbon
lifestyles easier and more possible
—angelaelizabeth, Redlands Bristol CRAG Web site

Business does know how to do this.
—Russell Mills, Dow Chemical at
UNFCCC COP (December 2007)

The elegance of the multilateral approach to climate governance is plain to see. A
legally binding global treaty engages all nation-states in common (and hopefully
enforceable) purpose. In theory, there is a smooth vertical development of policy that
draws on the accepted and traditional authority of nation-states, both in constructing the international treaty and formulating national regulations. International law
translates to national regulation, which directs domestic actions at more local levels.

Climate governance experiments are messier. They engage multiple actors in
less obviously common purpose. They challenge the known and comfortable
assumptions of centralized nation-state authority.[1] What's more, they seem to

be everywhere, simultaneously local and global. If the world fails to effectively address climate change, it will not be because of a dearth of ideas or activities geared toward solving the problem. On the contrary, everyone seems to be in the business of climate change. As the first two chapters make clear, actors of all sorts have recently begun engaging with climate change, making and implementing rules to shape climate behavior at all scales, and devising innovative ways for governing this problem that have little or nothing to do with negotiating international treaties. My claim is that in order to grasp whether and how climate governance experiments are likely to have an impact on the broad global response to climate change, we must be able to explain both the individual decisions to experiment (or at least how they became possible) *as well as* the emergence of patterns and relationships in the collection of experiments.

Doyne Farmer, a physicist, has said that at its heart, "science is about the telling of stories that explain what the world is like, and how the world came to be as it is."[2] My purpose here is to develop an analytic story that explains the observations described in chapters 1 and 2 and that will also provide insight into what we can perhaps expect from experiments (individually and collectively) and their broader potential to influence the global response to climate change (or not).

The goal here is not theory testing—positing an abstract explanation for various aspects of climate governance experimentation and ascertaining how well my explanation stacks up when compared with a series of alternative explanations. Although that is a valuable method of social scientific inquiry, the task at hand in this book requires a different approach, especially given the recent and multilevel nature of the experimentation phenomenon. Instead, this chapter is about sense-making—it seeks to construct an account of experimentation that makes sense of the two levels of innovation evident in the initial chapters—individual decisions to experiment and patterns in the collection of experiments. In other words, I want to develop an analytic framework that facilitates "thinking" experimentation.[3]

A Dose of Jargon: Experimentation as Complex Adaptive Social Construction

Developing the framework requires a brief turn to theory because it provides the plotline, if you will, of the climate governance experimentation story. We have to know something about how actors involved in climate governance might or should think and act, and theory helps us to see dynamics and events in new ways and to generate insights that are unavailable by just describing "what's going on."[4] I draw on a combination of social constructivism found in the international relations literature and insights from interdisciplinary complexity theory approaches for this task because of their focus on the social/relational side of global climate governance and on the ideas that provide meaning and guidance to actors engaged in addressing climate change.[5] Experimentation challenges conceptions of the nature of the climate change problem and proper responses to it. These theoretical perspectives are sensitive to the role of ideas and practices, as well as the importance of strategic interaction and are thus appropriate for this study.

Crucially, both social constructivists and complexity theorists concentrate on how actors connect with and are embedded in their broader context or environment, their communities or society—the broader world around them. In this book, the broader environment of interest is the governance context—the social norms, institutions, resources, and shared ideas that determine which actors get to make the rules, how rules will be made, and the general contours or substance of the rules. Both constructivism and complexity theory posit that a feedback dynamic connects actors and this governance context. This allows me to develop a framework that captures how a dominant mode of governance like megamultilateralism could be eroding and potentially giving way to new conceptions of governance embodied in the collection of climate governance experiments described in chapter 2.[6]

Paraphrasing John Ruggie, at any particular point in time, actors know what counts as governance.[7] They understand their governance context. This knowledge enables them to place themselves in the world and define their interests and desired actions.[8] The governance context thus informs and constructs actors—telling them who they are and what they want. For instance multilateral treaty-making in general came to be a dominant mode of international interaction in the 19th and 20th centuries.[9] This governance context shaped nation-states' identities and internal structures (e.g., seeing themselves as subjects of international law, developing bureaucracies to facilitate negotiating treaties, training international lawyers, etc.) as well as their goals and actions (consistently turning to international treaty-making to deal with international disputes).

In turn, the governance context is not merely a set of external constraints that actors face. It is created by actors' actions and interactions, which serve to reinforce or change shared ideas about who makes the rules, how they get made, and what they are. Actors' practices construct the governance context itself, even as that context shapes what those practices are. Multilateral treaty-making, for instance, only became a dominant feature of the governance context because states began making treaties in increasing numbers in the late 19th century, a trend that continued through the 20th century. As Denemark and Hoffmann observe:

> The emergence and early success of multilateral treaty- making alters
> the social context and states respond with changes in their expectations

about international interactions. The new expectations make further treaty-making more likely, reinforcing these early actions and reify-ing the appropriateness and utility of multilateral treaty-making. Eventually, this positive feedback lends multilateral treaty-making a taken-for-granted quality across large swaths of the global system.[10]

Figure 3.1 is a (simplified heuristic) visual representation of this dynamic. Moving along the arrow from agents to the governance context, the things that actors do—their practices, words, and beliefs—create, reinforce, or change shared understandings of what counts as governance (who gets to make the rules, authoritative actors, and what modes will be employed to make the rules). Moving along the arrow from the governance context to the agents, shared ideas about what counts as governance serve to condition what actors see as possible and plausible.[11]

However, this conception that actors shape their context, which in turn constitutes the actors, would seem to be a recipe for stability—a self-reinforcing dynamic. It often is, but both constructivism and complexity theory posit that this feedback relationship can produce periods of both stability and instability. When there is a stable governance context, most actors know which actors will make the rules and the kinds of rules that will be made, at least in general terms, and plan or strategize accordingly. In such circumstances, actors' actions reinforce the governance context. In an instable governance context, there is uncertainty about appropriate and effective rules and the means of making them. This uncertainty leads to novel actions that serve to erode old and create new governance contexts. At the core of both constructivism and complexity theory is the notion that stability and instability are phases or cycles through which social systems proceed (fig. 3.2).[12]

When the system is stable, then the feedback process captured in figure 3.1 is reinforcing a particular governance system. This is the state of most systems

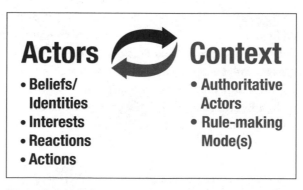

Figure 3.1   Feedback between Actors and Context

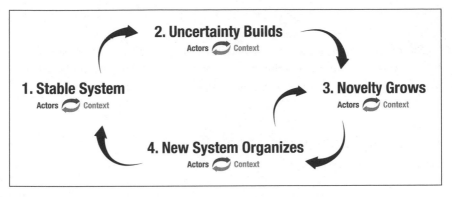

Figure 3.2 Transformation in Governance Systems

most of the time. Technological systems, policy paradigms, and governance contexts tend to lock in around particular technologies, policies, or configurations, and stability is the norm.[13] However, uncertainty inevitably builds in a system, no matter how stable—actors question or have diverse interpretations of the appropriateness or utility of the governance system, and they can even make mistakes and try something they shouldn't. Even a stable governance context will not and cannot be perfectly reproduced at all times. In fact, complex systems are inherently "characterized by a kind of uncertainty that presents opportunities for speculative acts of innovation."[14] A classic example of this from technology studies is how VHS won out over Betamax in video technology and became dominant (stable system), before being replaced again by DVD technology (a new stable system). Even in stable systems, there is enough uncertainty to catalyze innovative thinking. This uncertainty can emerge endogenously if actors within the system begin to contest the rules or exogenously through the introduction of new information or shocks to the system.[15] To be sure, actors are not passive. Both constructivists and complexity theorists consider that all political actors have an adaptive nature, reacting to what they see and experience continually updating their understanding of their context. They react to their own evaluations of what is happening around them and to new information making uncertainty in the feedback process inevitable.

This uncertainty catalyzes the production of novelty in the system—new ideas about governance, for instance. There is a subloop of feedback here because when new ideas about governance emerge, they tend to enhance the uncertainty in the system that gave rise to them in the first place.[16] Novel ideas about governance, when enacted, serve to construct and organize the governance system—eroding the old and organizing the new.[17] When the novel becomes normal, the system, and its new governance configuration, is stable again.[18] The key is to remember that the feedback loop depicted in figure 3.1 operates continually through the transition cycle visualized in figure 3.2 (which is why the feedback loop is pictured at each stage).[19]

So this is the plotline. When the details of climate governance are added, this basic framework illuminates the emergence of patterned experiments and generates expectations for the development of an experimental system.

Expecting Experimentation—From Stability to Uncertainty

Stability in the governance context results from centripetal forces whereby actors' beliefs/actions and the governance context are mutually reinforcing. Through the 1990s, just such a unifying, *centripetal* process was at play in the response to climate change, as all or most climate related activities were oriented toward the multilateral process. With few exceptions, NGOs worked toward convincing states to take action in the negotiations.[20] Cities and provinces developed very little in the way of climate policy.[21] Individuals and corporations looked to their states to take action and urged either restraint or boldness depending on how they understood the urgency of the problem. While the *content* of international negotiations that produced the UNFCCC and Kyoto Protocol was vociferously debated, megamultilateralism as the *means* of governing climate change was widely taken for granted.[22] By attending the negotiations, and orienting their discourses and practices toward this mode of governance, states, NGOs, multinational corporations, and even publics and academics reinforced the idea that the megamultilateral process was the way to respond to climate change (see fig. 3.3). Their actions and beliefs constructed the governance context in a manner that made it clear that the response to climate change was best effected by nation-states negotiating an international treaty.

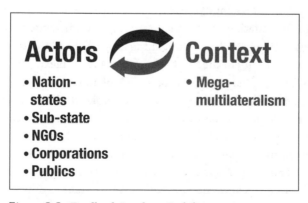

Figure 3.3 Feedback Leads to Stability

Yet the data presented in chapters 1 and 2 demonstrate that conventional wisdom is changing and that climate governance is no longer solely a matter of multilateral negotiations and national implementation. Actors began to think and act differently and consequently the feedback process is shifting into a *centrifugal* dynamic that is fragmenting the governance context. Instead of a central focus, multiple ideas about what counts as governance have emerged. Uncertainty has grown in the system, with potentially significant consequences for how the world will respond to climate change.

Transitioning out of stability—from a centripetal to a centrifugal dynamic— is not a simple matter. Constructivists stress that the shared ideas and material resources that comprise a stable governance context are real, consequential, and not easily altered.[23] The reinforcing of governance contexts (stage 1 in figure 3.2) is usually the dominant dynamic. Indeed, "critical junctures" where a stable context becomes instable are relatively rare.[24] Yet change is possible, and there is nothing inevitable about megamultilateralism as a response to climate change.[25] Constructivism and complexity theory both say that uncertainty is inherent in social systems—actors are always at least *potentially* uncertain about the governance context within which they exist and act. Notions of who authoritative actors are and what should be done are shared ideas that emerge from the actions and interactions of actors, so the rules of the game are potentially flexible and can change when beliefs, actions, and interactions change. As adaptive actors evaluate their actions (or the actions of accepted "governors") and their understandings of their context evolve, this can potentially lead actors to challenge existing governance structures (after negative evaluation). It is possible for actors to stop enacting megamultilateralism even when their beliefs and actions have been influenced by a governance context that is dominated by megamultilateralism. Actors are never fully conditioned by the governance context—they not only enact the governance context but also evaluate governance outcomes and reflect on the governance context.[26]

Thus, while in many cases actors go about their business via habit,[27] or through automatic social responses that are mostly "unreflexive and inarticulate through and through,"[28] because they ultimately have an adaptive nature, political actors are able to evaluate and react to governance outcomes and update their understanding of the governance context.[29] They are never so locked in to a particular governance context that they cannot change their behavior to enact other principles. The question is what conditions led some actors to conceive of acting differently in the global response to climate change—to stop reinforcing and enacting megamultilateralism and instead think differently about the global response. Ultimately the decision to experiment with climate governance in any particular instance results from specific concerns and circumstances of specific actors, so there are 58 stories of experimentation in this study that may all be different. That said, it is possible to identify enabling conditions that increased uncertainty around megamultilateralism, enhancing the possibility for experimentation and hence novelty to emerge in the governance context in the manner captured in chapters 1 and 2. Specifically, trends in the broader context of *global* governance, the signing of the Kyoto Protocol, and recent failures of the UN negotiating process all served to enhance the probability of experimentation with climate governance.

FRAGMENTING AUTHORITY

The existing megamultilateral process is certainly not the only, or even the most important, influence on how actors (cities, corporations, subnational governments, NGOs, or nation-states) conceive of political authority or the appropriate way to approach transnational environmental problems. There is a broader context beyond the issue of climate change that influences the way climate governance itself is conceived (fig. 3.4)—general notions of political authority and proper modes of governance that span issue areas, an example of which is how multilateral treaty-making as a generic governance tool permeates the international system.[30] One particular trend in the broad governance context noted time and time again by academic and media observers is that globalization—the increasing pace and volume of global flows of people, information, ideas, and money—has served to erode of the competence and authority of nation-states (individually and collectively) with profound effects on the way in which a range of actors see themselves and thus on their relationship with the dominant megamultilateral governance context in climate change. Cities, states/provinces, corporations, and more have begun to see themselves as authoritative actors in general, and this translates into an enhanced proclivity to see themselves as authoritative actors in climate change.

Conventional understanding of world politics holds that nation-states are the architects of the international system because they represent the highest level of political authority—they are sovereign. From this perspective, nation-states are the only actors that can truly manage transnational issues

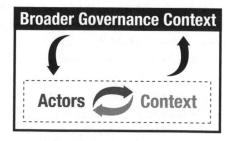

Figure 3.4 Embedded Climate Governance

because they can make legally binding rules enforceable domestically. The negotiation of consensual treaties and development of international organizations
that carry out nation-states' agendas is the natural (and some would say only)
way for sovereign actors to manage the international system. In this way, the
broader context of international relations—where multilateralism is a key
mechanism for sovereign states to interact—should reinforce megamultilateralism as the governance model for climate change.[31]

We are, however, in the midst of a general global shift toward the fragmentation of governing authority.[32] The ability of nation-states to manage transnational issues and the ability of nation-states to command authority, what
Rosenau calls the "readiness to comply with directives,"[33] have both been compromised by globalization. Saskia Sassen argues that "a good part of globalization consists of an enormous variety of micro-processes that begin to
denationalize what had been constructed as national."[34] Flows of knowledge,
people, money, products, and pollutants increasingly move globally and quickly.
This, in turn, has engendered a trend for actors other than nation-states to take
on the role of "governors" of transnational issues.[35]—political authorities able
to design and make rules themselves, rather than merely complying with the
directives of nation-states or the results of cooperation among nation-states
(treaties and intergovernmental organizations).

Scholars from multiple perspectives invoke such terms as *global civil society*,
global public domain, or *private authority* to convey the sense that the authority of
the state, and consequently that of interstate institutions (like megamultilateral
treaty-making), is eroding and/or diffusing to other levels of politics.[36] "Political
authority is being reconfigured in terms of both the entities of authority and the
modes of legitimation."[37] This tendency has been observed in multiple global
issue areas. Timothy Sinclair has tracked the activities and influence of private
bond rating agencies that have played a significant role in governing the global
financial system (with recently disastrous outcomes).[38] The rise of private security companies and the role that they play in conflicts and postconflict situations
all over the world is a growing concern in security studies.[39] In development,
scholars identify a clear trend of NGOs implementing development programs
and dispersing foreign aid.[40] The global environment has been especially fertile
ground for the emergence of authoritative actors beyond the state. Multiple
studies have demonstrated the role of cities, states, NGOs, and corporations in
setting standards for sustainable forestry, pursuing climate change activities,
and governing the trade in hazardous wastes, to name but a few that served as
part of the inspiration for this study.[41]

The increased "downloading" of responsibility for managing a range of issues
from nation-states (individually and collectively) to other actors[42] happens consciously at times as nation-states shed responsibilities and unconsciously at
others as nonstate actors seize the authority to act. Regardless of the source, the

erosion of the authority of nation-states and state-centric governance approaches makes it easier for a variety of actors to envision themselves as authoritative designers of rules for responding to climate change. This is a background condition that shapes how adaptive actors understand climate governance. It opens up the possibility of a shift in which actors are considered authoritative and consequently the possibility of conceiving new means of responding to climate change.[43]

THE LEGACY OF THE KYOTO PROTOCOL

The signing of the Kyoto Protocol in 1997 ironically served to enhance the possibility for experimentation and erosion of the dominance of megamultilateralism. The Kyoto Protocol made clear the challenge of reducing carbon dioxide emissions at various scales and, as important, the potential profit to be made, thereby motivating innovation. An example of this catalytic process was the emergence of the Greenhouse Gas Registry (now the American Carbon Registry). When Environmental Defense and the Environmental Resources Trust observed the importance of market mechanisms like emissions trading in the Kyoto Protocol negotiations (an option they strongly advocated throughout the negotiation process), they identified a need for infrastructure. A significant part of this task was to get companies ready to calculate their carbon footprints, as well as measure and account for emission reductions that would eventually be traded among regulated entities. They took it upon themselves to devise rules to guide corporate measurement and trading by launching the Greenhouse Gas Registry in 1997.[44] Crucially, this governance experiment was not directly tied to the treaty process. It was instead a set of civil society actors seizing the authority to shape the response to climate change. Similar awakenings to the importance of climate change motivated experimentation in cities, corporations, and carbon markets.

Experimentation began slowly. From 1997 to 2001, only a few experiments emerged while the international community negotiated the details of implementing the Kyoto Protocol. Moreover, these early experiments were anticipatory, building the infrastructure to implement the climate actions promised in the agreement. For instance, most of the emissions trading venues that emerged in Europe (UK and EU) immediately after the Kyoto Protocol was signed were developed, at least in part, to gain experience and prepare for what was intended to be a global cap and trade system.[45]

Yet while the experiments that emerged in these early years were working in anticipation of a functioning global treaty, their activities institutionalized the possibility of governance mechanisms outside the dominant megamultilateral process, and in so doing sowed seeds of uncertainty in the governance context. This is the exact dynamic that Sassen has discussed. Even when the multilateral process succeeds, it produces activities that undermine the stability of that governance mode.

As other actors prepare for implementation—making rules and starting up climate activities—that enhances their claims to and self-conceptions of authority over the issue. The CCP program is a quintessential example of this. It emerged on the heels of the 1992 Earth Summit and signing of the UNFCCC in order to help cities get ready to implement climate policy. Enhanced municipal authority over climate change[46] ultimately created some degree of uncertainty in the multilateral climate governance context, at least for participating cities that could now see themselves as governors of climate change. Now, if the Kyoto Protocol had turned out to be an effective agreement, engaging the major players and serving as the focal point for a stringent multilateral climate regime, extra-Kyoto mechanisms would likely be viewed as part of the implementation process and only significant as evidence of the need for wide buy-in from a range of societal actors necessary to fully address climate change. The centripetal cycle would still have dominated even as other actors began to work on climate governance. Such was not the case.

STALEMATE

The Kyoto Protocol process moved steadily toward stalemate in the late 1990s, exacerbated by the withdrawal of the United States in 2001, a condition in which the Kyoto process remains mired today. The failure the 2009 UNFCCC COP negotiations in Copenhagen is only the most recent, and perhaps most spectacular, reminder of just how difficult it is to negotiate a global treaty for an issue as complex and contentious as climate change. It is likely that Copenhagen will be viewed as a turning point, not in the direction of successful international treaty-making but as the first step toward a significant rethinking of the use of megamultilateral processes for addressing climate change.

This stalemate provided a third, and perhaps the most significant, source of uncertainty around megamultilateralism. When actors call into question the legitimacy and efficacy of the prevailing governance model, the uncertainty in the governance context, already present because of globalization, increases. As noted, dismissal and critiques of the Kyoto Protocol grew in the late 1990s and early 2000s. The negative reaction to the Kyoto process from actors and observers increased the uncertainty surrounding the dominant megamultilateral governance model as the accepted procedure for pursuing the global response to climate change.

A major aspect of increasing uncertainty through stalemate was obviously the U.S. withdrawal from the Kyoto Protocol in 2001. As the largest contributor of absolute greenhouse gas emissions (until roughly 2007, when China surpassed the United States), the United States was and remains perhaps the single most important nation-state in the multilateral response to climate change. The U.S. withdrawal threw the multilateral process into a tailspin. While the Kyoto Protocol came into force in 2005, it never generated the necessary implementation momentum, with the exception of the European Union's efforts and aspects of the CDM, to really address climate change effectively. Actors reacted strongly to both the U.S. withdrawal and the overall malaise in the UN negotiations. In one particularly visible example, the writing and signing of the U.S. Mayors' Conference Climate Protection Agreement was an outright repudiation of the U.S. withdrawal from the Kyoto Process.[47] More broadly, both the U.S. withdrawal and the stalemate that characterized most of the 2000s finally convinced many actors that the megamultilateral process was broken or at least that momentum would need to come from outside the process.

Choosing to Experiment: From Uncertainty to Novelty

Perhaps more than any other scholar of world politics, James Rosenau has identified and considered innovation in transnational political communities. He notes that "the emergent epoch is one in which communities are being reimagined" and that "the number of new entities that draw people . . . seems to proliferate as the complexity of the emergent epoch deepens."[48] In climate change, the combination of enabling factors resulted in a range of actors becoming less and less certain that megamultilateralism was the appropriate and effective way to address climate change. When uncertainty goes up, the potential for experimentation also increases.[49] A quick upsurge in experimentation subsequently resulted after 2001—mainly in the global North as a way to either move beyond what states had agreed to do (i.e., Europe) or to act in the face of national inaction (i.e., North America). The proximate causes of experimentation—the translation of enabling conditions into specific decisions to innovate—are diverse. While it is beyond the scope of this book to explain the specific motivations that drove adaptive actors to realize the potential to experiment in each of the 58 cases,[50] it is possible to describe classes of motivations evident in climate governance experimentation:[51]

- *Profit*. Carbon offsetting organizations, for instance, implicitly promote a governance model based on individual responsibility for climate change. This model is profitable for those who are selling offsets. Indeed, the business of environmental governance is increasingly lucrative, with the development of carbon markets, public-private partnerships to exploit genetic resources, and certification measures that provide certain products with visibility and a "green stamp" of approval.
- *Urgency*. A more standard explanation given for the motivation to innovate is that some actors desire to move quickly on environmental problems and are concerned about the slow pace of multilateral processes. Citizen groups and subnational governments, especially in those states like the United

States and Australia, where federal authorities were recalcitrant, are often characterized as being motivated by urgency.

- *Expansion of authority and claims on resources.* Urgency aside, we should be cautious in ascribing only altruistic motives to those entities that want to move quickly on environmental problems. While cities may be genuinely motivated by urgency in pushing for quicker action than is available multilaterally, it is crucial to note that they are pushing for *municipal* action. An event at the University of Toronto brought together two former mayors (Ken Livingston of London and David Miller of Toronto), who not only stressed the importance of cities in addressing climate change but also called for more resources to come to cities from other levels of government. Similarly at the 2007 UNFCCC COP in Bali at side event after side event, different actors (provinces, cities, corporations) stridently claimed the mantle of "most important actor" for dealing with climate change.[52] Actors may experiment with governance as a way to enhance their own authority and resources relative to other levels of political organization (state or nonstate).

- *Ideological expression.* One the one hand, the recent age of neoliberalism has brought a greater number of actors into governance experiments who believe, as a matter of course, that markets and private organizations are more effective and more efficient in the provision of governance functions (hence the liberal environmentalism observed in chapter 2). On the other hand, hosts of other civil society arguments articulate the view that NGOs and other nonstate actors (and processes such as "stakeholder" engagement and management) bring needed aspects of representation and legitimacy to transjurisdictional governance that more traditional state dominated governance models lack.

Driven by one or more of these factors, initiating actors decided to change their orientation to climate governance. Rather than (or in addition to) orienting their activities toward the treaty-making process, they began to create initiatives of their own. Novelty emerges in the climate governance system (fig. 3.5) and accelerates over time as initial innovations breed further uncertainty, which can subsequently catalyze additional novelty—the subloop in figure 3.2. Once the centrifugal cycle emerges, it has a reinforcing dynamic all its own. The question is what kind of governance context is emerging now that a fragmenting centrifugal cycle characterizes the actor-context feedback in the governance of climate change.

The Experimental "System" Emerges

To this point, then, the account of experimentation has followed the script laid out in figure 3.2, which indicates that novelty in a governance system is expected to follow an increase in uncertainty. Specifically, novel climate initiatives resulted

Figure 3.5   Novelty and Fragmentation of the Governance Context

from adaptation by multiple actors at multiple levels entailing a reframing of climate change and ultimately a seizing of authority over responses to climate change.[53] Since 2001, the actor-context feedback cycle has been churning away creating novelty in the global response to climate change. But, of course, the transition from centripetal to centrifugal cycle is not solely a dynamic of erosion. The experiments actors engage in also produce a new governance context that allows for different approaches to governing climate change and for actors to construct new ways of managing the global response to climate change. They are organizing a new system of governance, the form of which was evident in chapter 2—diverse, functionally differentiated experiments, bound together with a common liberal environmental ethos. The framework developed here illuminates why novelty has taken this form and what we can expect as the novelty normalizes—the experimental system's internal dynamics and interactions with the still relevant multilateral governance system.[54]

THE SELF-ORGANIZATION OF AN EXPERIMENTAL SYSTEM OF GOVERNANCE

The patterns observed in chapter 2 are expected from the perspective of the teachings of constructivism and complexity theory as the results of self-organizing dynamics:

> Complex systems are characterized by a capacity for self-organization, that is, the ability to rearrange and reform their patterns of organization in mutual adaptation to changing needs and capacities of their components as well as changing opportunities and demands from the environment.[55]

The components of the climate governance system (individuals, cities, states/provinces, NGOs, corporations, and nation-states) are in the process of "rearranging and reforming their patterns of organization." They are beginning to transform the novelty of climate governance experiments into a nascent experimental system of governance.

This is precisely why looking at the experiments collectively is so important. Because system-level characteristics, like the common liberal environmental foundation and the functional differentiation obvious in chapter 2, are emergent properties, we cannot predict what an experimental governance system might look like by examining individual experiments. Experimenters are reacting to and working on their "local" issues, but according to the perspective drawn on here, their individual actions create a larger system of governance. The patterns we see are thus the contours of the nascent experimental governance system. Organization theorists call this an organizational shadow that emerges when adaptive actors organize, "each following its own rules of behavior," in order to "acquire information, to learn, to conduct political activity, and to change the system they are a part of."[56]

In self-organizing systems we expect to see some key features and dynamics, especially the use of old tools/motifs used in new arrangements and the emergence of clusters that result from interactions and redundancies that arise in the system. The first of these are already evident in the data presented in chapter 2. A striking feature of the 58 climate governance experiments is how familiar they feel. Though it is easy to recognize that something new is going on, it is equally easy to recognize the kind of things the new initiatives are doing. Both constructivism and complexity theory contend that when actors innovate, they do not do so from scratch. The novelty that we see in the third stage of the transition cycle is not a complete break from what has come before. On the contrary, initiating actors draw on existing resources, both ideas and material resources when devising experiments.[57]

The common liberal environmental foundation across experiments should be thought of in this light. The idea that economic growth is compatible with environmental protection has pervaded discussions on environmental issues for the last two decades (at least). Self-organizing experiments are shaped by and draw upon this set of widely accepted, if not uncontested, ideas even while they employ them in new contexts and towards different ends. In addition, specific configurations of initiating and implementing actors as well as activities that experiments undertake are also familiar. Once the dominance of the megamultilateral governance model begins to erode, actors start playing with different pieces and motifs to make experiments. The novelty they create is thus familiar in many ways. Cap and trade is a generic tool that shows up in a number of experiments. It emerged in efforts to address acid rain in the 1990s, was included in the Kyoto Protocol, and has spread to diverse jurisdictions.[58] Smaller experimental multilateral arrangements like Methane to Markets and the Major Economies Forum follow the example of small group diplomacy seen in the G-8 and G-20 processes. The name "C40" is actually a deliberate reference to the G-20 (C40 began as C20). Voluntary corporate social responsibility initiatives spawn transnational collaborations among corporations. Carbon rationing action groups explicitly use the Kyoto Protocol as a model for their individual-based initiative, and the U.S. Mayors' Conference Climate Protection agreement pledged to meet Kyoto goals. Subnational governments who have cooperated in other areas, for instance over the Great Lakes in North America, set up new experiments to deal with climate change and draw on extant cap and trade mechanisms to respond to climate change. When we look at the collection of experiments, we see patterns, categories, and familiarity rather than randomness, in part because experimenters are drawing on existing themes (networking, planning, voluntary action, enforceable action) when they innovate, and these motifs in different combinations serve to structure the experimental system.

As the experimental system develops, we should also expect to see certain kinds of activity among experiments: networking, a combination of competition and cooperation between initiatives, the emergence of communities of practice, and the development of redundancy in the system facilitated by the functional differentiation noted in chapter 2.[59] Examples of these dynamics are evident with even a cursory look how climate governance experiments are interacting, offering some initial reason to think that these expected dynamics are emerging among climate governance experiments.

During the 2009 Copenhagen negotiations, C40 and ICLEI collaborated on a Climate Summit for Mayors (December 14–17, 2009) that essentially competed with The Climate Group's Climate Leaders Summit (December 15–16, 2009). It is not uncommon to hear participants (cities, states, corporations) stridently proclaim themselves "the key actor for addressing climate change" while they are also collaborating with other experiments and projects on various projects and initiatives. Collaboration, competition, and networking extend beyond the focal UN climate negotiations. The Climate Group, for instance, lists ICLEI and the Network of Regional Governments for Sustainable Development as partners and collaborated with the World Business Council for Sustainable Development in developing the Voluntary Carbon Standard for carbon offset quality and integrity. Registries and offset standard setting experiments like the Climate Registry, the American Carbon Registry, Voluntary Carbon Standard, the Carbon Action Reserve, and the Climate, Community, and Biodiversity Alliance have significant overlap in terms of people who have worked on their designs even while they serve different niches in the carbon market and compete to be recognized in both the market and among regulatory bodies.[60] The three major emissions trading systems being developed among U.S. states and Canadian provinces have begun meeting (initial meetings were in July and November

2009) to discuss both how to link the systems and how to best interact with the potential development of a U.S. federal cap and trade system currently being considered in the U.S. Congress.[61]

In general, then, we would expect the evident functional differentiation to foster this cooperation, competition, and redundancy among experimental initiatives. The different roles that experiments play can facilitate cooperation as experiments with different skill sets, goals, and governing modes have opportunities to work together. However, the limited differentiation (only 4 governance models) means that different experiments may play the same roles and potentially fill the same niches, which can lead to redundancy and competition in the experimental system. This is not necessarily negative—competition can serve to enhance innovation in the experimental system and redundancy that develops as multiple experiments work in similar areas can provide resilience for the global response to climate change.

These relationships develop and play out in specific areas creating activity clusters that also structure the experimental system. Self-organization is characterized by, for lack of a better word, clumpiness in a system. Rather than being randomly distributed or independent, components of the system will group together, "clustering around points of energy" or creating a "basin of attraction" around a particular function or activity.[62] The functional differentiation observed in chapter 2 facilitates this aspect of self-organization. Without external design, a (limited) range of functional roles have emerged in the population of climate governance experiments—with "different components perform[ing] different functions simultaneously."[63] The relationships and interactions that functional differentiation makes possible fosters substantive clustering like that hinted at in the anecdotes above. These substantive clusters are developing rapidly in the experimental system. Chapter 5 examines an experimental basin of attraction around deployment of climate-friendly technology in cities, while chapter 6 explores another one concentrated on the development of carbon markets. These clusters serve as the media through which the experimental system is organized and functions. The functional differentiation configures the kind of relationships that experiments can have (cooperative, competitive, parallel) and the niches they can develop. Substantive clusters provide specific sites of interaction that draw experimental activity and are the specific areas where experiments actually function and work to respond to climate change.

Viewed collectively, climate governance experiments are normalizing into a system of governance. But it is a nascent system in which organization and relationships have yet to fully emerge or stabilize. Whether this developing experimental system will continue on a path to coherence and whether this path will lead to an effective global response to climate change are still open questions explored in the chapters that follow.

THE GOVERNANCE SYSTEM NORMALIZES

The story of climate governance experimentation does not imply that megamultilateralism will disappear or that it is no longer important. On the contrary, it is crucial to understand the role expanding experimentation will play in the overall global response to climate change because multilateral treaty-making is still a significant focus of international effort. Fortunately, the same perspective that helped tell the story of experimentation to this point facilitates making conjectures about how the experimental system will develop and how it might interact with the multilateral system.

One way to conceive of the emergent organization evident and expected among governance experiments is that a division of labor is emerging in opposition to the centralized, but stalemated, megamultilateral process. There is plenty to do in responding to climate change, and by specializing and breaking down the tasks into manageable pieces, perhaps a more effective decentralized global response will emerge.[64] Already within a decade of the emergence of experiments, we see this kind of developing specialization in the functional differentiation described in chapter 2. Infrastructure builders like registries and investor networks provide the information necessary to make decisions about carbon and measure the changes in greenhouse gas emissions. Networkers make sure that best practices and the most up-to-date information about technology can spread quickly. Voluntary action and accountable action experiments draw on the foundation set by other experiments to proceed with specific cuts to greenhouse gas emissions (tailored to their particular level of political authority). The relationships and interactions made possible by functional differentiation could lead to efficiencies and effective means of generating friction and disruptive change in the larger global response to climate change.

The development of voluntary carbon markets, discussed in detail in chapter 6, is a quintessential example of this dynamic. Climate registries or disclosure initiatives (the Climate Registry, the Carbon Disclosure Project) and offset standard setters (the Climate Action Reserve, the American Climate Registry) are crucial parts of the infrastructure necessary for carbon markets and emissions trading systems to function. A whole ecosystem of infrastructure builders is emerging so that actors can measure, account for, offset, and trade emissions (see chapter 6)—in both the experimental world (the accountable actors model is dominated by emissions trading initiatives) and the more conventional climate governance world (the trading system set up by the Kyoto Protocol and/or EU). Additional linkages between experiments are also emerging around cities and technological deployment. The Climate Group (networker), for example, has formally linked with the Connected Urban Development program (voluntary actor) in an effort to scale up the latter's activities through the network of cities in The Climate Group (see chapter 5).

Yet the division-of-labor path is not the only conceivable way that the experimental system may develop. Different experiments may fulfill different roles
in the experimental system, but it could be a far from efficient or effective specialization. Experiments could generate substantive clusters in a way that
works against disruptive change. Instead of a division of labor, we could see
Tiebout sorting.[65] The Tiebout model was originally developed to explain how
competing localities or parties looking to attract citizens/members from a het-
erogeneous society would offer different amenities/platforms and serve as a
sorting mechanism. People would move to locales or join parties that matched
their preferences.[66]

There is the potential for a form of Tiebout sorting to develop in global climate governance. With the opening and fragmenting of climate governance,
actors are able to create and/or join experiments that suit them and their preferences best. They can strategize, asking what is best for me materially and/or
what is appropriate for my values. Tiebout was making claims about democratic
accountability—the more effective an institution is at sorting, the better represented people will be. However, in the case of climate experimentation, this
sorting action may be detrimental. Actors may find just the right kind of experiment that suits their needs and values. The United States was at the forefront
of pushing voluntary, small-group multilateralism during the Bush administration because it fit their interests in moving slowly on climate change. Every
actor may find an experiment to suit its interests, but this does not necessarily
equal an effective response to climate change. The question that remains is
whether open sorting into experiments that match actors' preferences will provide enough of a response. If experimental sorting occurs in the absence of
legally binding enforcement of broader climate change goals and activities—
enforcement that can likely only be achieved through international treaties and
national laws—it may not provide an effective catalyst for climate action. If substantive clustering emerges in a way that allows experimenters to remain comfortable, just doing what they have always done, disruptive change will be
difficult to achieve through experimentation. Friction will not develop and
experimentation will not catalyze broader change.

The way the experimental system evolves, into either a division of labor or a
Tiebout sorting process, is a crucial question, because its development shapes how
the experimental and multilateral systems will interact. Two extreme outcomes
seem implausible, but are worth noting. The first is that experimentation could
simply recede in the face of resurgent multilateral activity or nation-state action.
This seems unlikely, both because the multilateral treaty-making is currently
mired in a deep political ebb and because it is becoming increasingly plausible
that megamultilateral treaty-making might simply be unable to overcome the collective action problems that have plagued it for the last two decades. In addition,
the momentum behind experimentation is growing even as the multilateral

negotiations flounder.[67] The second is that experimentation will eclipse and replace multilateral treaty-making. This is unlikely because ultimately some enforceable rules will likely be necessary at a broad scale to effectively address climate change.

The most likely scenario is a middle ground where the experimental system and the multilateral system will co-develop, at times in parallel and at times interacting. This would not be a diminishing of the significance of climate governance experimentation. On the contrary, the center of gravity in the global response to climate change may have already shifted toward the experimental system, and it may become the driver of the global response even while multilateral treaty-making remains a tool of climate governance. If the experimental system develops into a functioning division of labor, it could set the agenda for national and global action on climate change, creating both the friction that catalyzes broader action and the infrastructure to smooth the path to a broader response. Ironically, the more the experimental governance system develops as if it is an alternative to the traditional multilateral approach, the more likely it is to be a driver of effective multilateral treaty-making and national action and avoid the pitfalls of Tiebout sorting. Done well, climate governance experiments may be just what the global response to climate change needs. Climate governance experiments are innovative. They are pushing the envelope of what is possible. They are actively seeking out and creating gaps in the response to climate change and attempting to fill them. They can undertake activities that raise issues about and create interests in finding broad solutions to climate change. They can generate the human and organizational capacity for responding to climate change that can be used in the service of a broader response.

Conclusion

The skeletal essence of constructivism and complexity theory embedded in figures 3.1 and 3.2 helped structure a story of climate governance experimentation. The story is incomplete, however, because climate governance itself has yet to fully move through the transition cycle visualized in figure 3.2. It is impossible to give a full account of how the novel becomes normal in the case of climate governance experimentation precisely because we are still somewhere between stages 3 and 4 of figure 3.2. Yet we now have a way to think through and about experimentation and the agenda for moving forward is clearer. First a plausibility check is required for the account developed here. Chapter 4 examines vignettes of experiments from each of the 4 governance models discussed in chapter 2 to see if the story developed to make sense of the collection of experiments holds up when examining specific experiments more closely. Can we see the motivations to experiment and the developing experimental system when we look closely at individual experiments? Chapters 5 and 6 then examine clusters in the emerging experimental system in order to see if the expectations about self-organization are plausible and to begin the process of critically assessing the experimental system's development and interactions with the multilateral system.

4

Experimenting in Practice

> We're going to do what we're going to do regardless [of the
> Copenhagen outcome]. . . . We can't solve this problem alone.
> —David Cadman, chair, *ICLEI (Copenhagen, December 2009)*

O Brave New World That Has Such Experiments in It

A new experimental climate governance system sounds great. It provides hope
in the face of growing despair over the fate of global treaty-making.[1] In early
2010, the United States had to urge China, India, and Brazil to be more committed to the Copenhagen Accord in order to avoid it being "stillborn,"[2] even
while prospects for U.S. federal cap and trade legislation (and climate legislation
more generally) dimmed during the subsequent summer and fall of 2010. The
momentum behind other negotiating tracks toward a legally binding treaty is
uncertain at best. Even the Intergovernmental Panel on Climate Change, the
once heralded winner of the 2007 Nobel Peace Prize, is under sustained attack
for a series of missteps that have given climate change skeptics ammunition to
question (mostly without significant merit) the scientific basis of climate policy.
With the obstacles to U.S. federal action and multilateral treaty-making growing,
rather than receding, it is easy to take some solace in statements like those from
Governor Charlie Crist, a Republican from Florida: "What they're [states] doing
is already genuinely significant. . . . You have thirty-three states with climate
plans. These aren't just vague aspirational plans like you saw under the Kyoto
Protocol, but concrete goals on efficiency, renewables—tangible things that are
being written in law."[3] The experimental world seems to be the only part of the
global response to climate change that is moving forward. However, the quotation

from David Cadman in the epigraph to this chapter is telling. Experimentation moves forward even in the face of failure in Copenhagen, but there are major questions about whether experimental governance initiatives will be enough. Critical assessment of experimentation is necessary to see if and how it might live up to its promise.

To this point, the analysis has been telescopic—examining the world of experimentation from afar and in broad strokes. From this perspective, it seems plausible that experiments emerged through a transition in the governance context (that they helped further) and that an experimental system of governance is emerging in a self-organized, organic fashion that has some coherence. This chapter is more fine-grained, demonstrating that the analytic story derived to explain the broad dynamics of experimentation also holds when examining individual examples of experiments from each of the four governance models—networkers, infrastructure builders, voluntary actors, and accountable actors. Examining canonical examples of functionally differentiated experiments—their histories, activities, and interactions—imparts a concrete analysis of the process of experimentation while also lending insight into the developing structure of the experimental world.[4]

The Climate Group: Networking Corporations and Subnational Actors

While officially launched in 2004, The Climate Group had its origins in a May 2003 "Conference of the Reducers," a title with pointed similarity to the UN COPs.[5] This meeting in the Hague was sponsored by the German Marshall Fund, the Center for Clean Air Policy, the Rockefeller Brothers Fund, and the Netherlands National Institute of Public Health and the Environment.[6] Business leaders and officials from subnational governments were invited to tell their stories about what they were doing for climate change and how it was good for the environment, their bottom lines, and citizens. What was remarkable to participants was how little they knew about each other before this meeting. In the years following the U.S. withdrawal from the Kyoto Protocol in 2001, there was a dearth of good news about climate change, and "getting the message out that acting on climate change was good as opposed to a job killer was not easy"; the fact that some corporate and governmental actors were taking significant actions was "an untold story."[7] Buoyed by this meeting, Michael Northrup of the Rockefeller Brothers Fund approached Steve Howard, the now CEO of The Climate Group, to develop an organization that would not only get these stories told, but put reducers in contact with one another. "Participants at the May reducers meeting felt that reductions could be accelerated by creating mechanisms for sharing practical experiences. Broadcasting positive success stories, they felt, could offer greater confidence to policymakers with respect to the practicality of reducing emissions cost effectively."[8]

The idea of an organization dedicated to bringing together like-minded "reducers" also appealed to Prime Minister Tony Blair, who was looking to move climate cooperation forward in the wake of his closest ally's dismissal of the issue and to include climate change as a central part of the agenda for his chairmanship of the 2005 G-8 meeting.[9] Blair was looking for a way to work internationally on climate change without running into the problems faced in the multilateral negotiations, and he wanted to make an economic case for addressing climate change, arguing for the "importance of a Climate Group that involves not just states and cities but also business so that there are practical, clear examples of how good environmental policy is also good business policy and is right for growth."[10]

The Climate Group was officially launched in April 2004, with original members hailing from both the governmental (e.g., Germany, California, Connecticut, state of Victoria, London) and corporate (e.g., British Petroleum, HSBC, Shell, Lafarge) worlds.[11] From the outset, and even from the activity in the year leading up to the launch, the dual effects of the Kyoto Protocol discussed in chapter 3—its signing and the stalemate—were palpable. Comparing the statements from Tony Blair at the launch of The Climate Group and a description of the initiative from the Rockefeller Brothers Fund captures these mixed motivations:

> We are committed to the Kyoto Protocol. We believe it is essential that we have that implemented. We in our country will abide by our Kyoto targets, but I just want to make one point to you. When I asked for an analysis to be done by David King and his colleagues of what the true scale of the challenge was, *we learned that even if we were to implement the Kyoto Protocol, it falls significantly short of what we will need over the next half century if we are to tackle this problem seriously and properly.* So even, and this is a tall order in some ways at the moment, if we succeed in getting support for the Kyoto Protocol, we are still, even having done that, only in the position of having achieved a first step. It will be an important recognition, but it is only a first step and we need to be building a clearer understanding of the fact that even with Kyoto we are still a long way short of what we actually need to do. And we've got to build support in the political institutions of which we're a part in order to make sure that that case is properly understood. I think we have to make sure that this occupies, as an issue, a central place in political decision making beyond any election or parliamentary cycle. It's beyond the life of any government. It's beyond the life of any passing political phase. It has to be there, central in the politics of each country, built up

not just from support within government, but from support within civic society over a period of time. (Tony Blair)[12]

At the time of The Climate Group's launch in 2004, international action on climate had slowed down. *The United States had slammed the door on Kyoto, and the future of climate leadership looked uncertain.* But The Climate Group's founding members, including Marks & Spencer, HSBC, British Petroleum, and Shell, were determined to get the ball rolling and move well beyond Kyoto's modest goals. And they offered a novel approach to get there. (Rockefeller Brothers Fund)[13]

Implementing actors, both subnational governments and corporations, have more specific interests in joining The Climate Group. As Callum Grieve, director of External Affairs North America, noted, The Climate Group "became the sweet spot between business and government."[14] This was not about altruism or addressing climate change solely for the sake of solving the problem—though The Climate Group is an experiment that appeals most to actors motivated by urgency and committed to progressive action on climate change. Both corporate and subnational actors perceive value added from the networking practices put in place. Certainly the overall goal articulated by Steve Howard at the launch of The Climate Group was relevant. He suggests that The Climate Group was the

right idea at the right time. We know that there are many leading companies and governments around the world dedicated to reducing their emissions. By bringing the key players together we believe we can accelerate progress and avert dangerous climate change.[15]

Implementing actors that are interested in doing so draw on The Climate Group network to meet their own needs. For instance, the network provides subnational actors with corporate endorsement when they want to move ahead on climate change, allowing them to deflect the critique that addressing climate change will cost jobs or hurt the economy. As Jane Gray, former director of the cities, states, and regions program put it, "when sub-national leaders want to do something progressive on climate change, having corporate leaders make the point that taking action was good for the climate and good for business is a huge boon."[16] A key example of this dynamic emerged around the signing of Assembly Bill 32 in California. In addition to measures like mandating greenhouse gas reporting and identifying early reduction actions, this law signed in 2006 commits California to reduce greenhouse gas emissions to 1990 levels by 2020.[17] In the lead-up to the signing of the bill in September, The Climate Group convened a July energy roundtable with "a select group of prominent CEOs and business leaders from leading California and international companies . . . to share ideas on how business and government can work together to reduce greenhouse gas emissions."[18] Major corporate leaders that attended the roundtable also backed the new law, including energy heavyweight Pacific Gas and Electric, making it that much easier for California to move ahead on this ambitious state-level initiative.[19]

For corporations, The Climate Group's status as a relatively independent voice on climate change provides legitimacy for corporate climate activities and a shield against accusations of "greenwashing."[20] Sarah Skikne, a corporate engagement manager for The Climate Group, notes that in addition to the value in sharing information, corporations enjoy the stamp of approval of their climate change activities from a respected third party.[21]

The network also facilitates moving beyond corporate social responsibility initiatives, beyond pilot projects into actual deployment of climate friendly technologies.[22] Phil Jessup, a founder of ICLEI's CCP and the C40 group, is currently working to scale up an LED lighting project in cities across the world.[23] This may seem like a relatively small initiative, yet The Climate Group notes that lighting accounts for 10% of global greenhouse gas emissions.[24] Street lighting is a significant percentage of this globally (8%), and cities own a good chunk of street lighting.[25] By networking municipal governments and corporations, The Climate Group has been able to facilitate a global pilot program to bring LED street lighting (50–70% lower emissions than traditional lighting) to major global cities (e.g., New York, London, Hong Kong, Mumbai, and Calcutta), engaging a dozen corporations that manufacture LED lighting.[26] Initial evaluations of the results from the pilot project in New York are showing reductions of up to 80%.[27] The path to scaling up this initiative is relatively straightforward:

> First we are conducting a global trial of LED lamps in large world cities
> to evaluate how newly commercial products actually perform. Next, we
> will work with these cities to scale up their pilots and further enlist our
> broader membership as early adopters. Finally, we will work with gov-
> ernments to encourage more friendly LED and smart control policies.[28]

Thus the networking that they provide has ancillary benefits beyond "broadcasting positive success stories"[29] and connecting like-minded actors. The connections and association with The Climate Group provide implementing actors with legitimacy and political cover to pursue climate activities individually and jointly. The connections have the potential to produce synergy on multiple dimensions—the members of The Climate Group appear to be doing more than just talking and exchanging information. While the day-to-day work of The Climate Group consists of sustaining member relations and gathering and sharing information, its actions raise the profile of climate action among corporate and subnational actors and pave the way for taking specific actions.

In 2007, The Climate Group's focus changed slightly, or perhaps more accurately it grew. Networking continued (and continues) apace, but policy advocacy on the global stage and technological deployment like the aforementioned LED program (others include electric cars, carbon capture and storage, smart grids, smart buildings in the cities, states, and regions program) became the key priorities for the organization. Networking has laid the foundation for these expanded activities,[30] and the transition is rather one of moving beyond solely networking— getting like-minded actors together—to networking with specific purposes driving the linkages.

But of course The Climate Group is not simply internally focused. It has significant links with other experimental initiatives and interacts with the multilateral process. In The Climate Group's external relations (as opposed to those with its implementing members), there is evidence of the development of the experimental system and the friction it can cause in the multilateral governance system. The Climate Group's relationships with other climate governance experiments span the cooperation-competition spectrum and are additionally creating redundancy in the experimental system of governance. Two examples illustrate this dynamic:

- The Climate Group participated in the convening of the C20 group in 2005—the organization that would eventually become the powerful C40 network of large cities.[31] Yet C40 and The Climate Group do approach (discussed further in chapter 5) technology deployment in cities differently. Phil Jessup notes that C40 prefers to let the market make choices, while The Climate Group has decided to pick a winner (at least in the lighting sector) and scale up pilot projects.[32]
- The Climate Group worked with the World Business Council on Sustainable Development (another networking experiment) to develop the Voluntary Carbon Standard in 2006–2007 as a way to bring some integrity to the voluntary offset markets.[33] But of course, this standard competes with those developed by other infrastructure building experiments like the Climate Action Reserve and the American Carbon Registry (addressed more fully in chapter 6). Mary Grady of the American Carbon Registry notes that there is both collaboration and competition among standard setters and that is good for the sector because it generates innovation in both project development (i.e., methodologies for producing carbon reductions) and standards (ways of verifying reductions).[34]

These are the very dynamics expected in a self-organizing system. The Climate Group plays a specific role—networking subnational governments and corporations—and this kind of activity makes for natural connections to other experiments that work with subnational and market-oriented actors. They both collaborate and compete with other experiments and actually recognize the value of both. Alison Hannon, Midwest regional manager, observed that competition is not always a bad thing and that collaboration with other initiatives has gotten much better.[35] In these cases, efforts to aid the launch of the experiment that would become C40 created two major technology deployment efforts in city networks, and multiple experiments began working to set standards for carbon offsetting independently. These efforts have the unintended consequence of increasing redundancy in the experimental system. As the experimental system develops, similar, but not duplicative, initiatives arise as initiating actors attempt and discover new modes of responding to climate change. In chapters 5 and 6 I will examine how this kind of activity creates the expected clusters of activity that structure the experimental system.

The Climate Group's efforts also contribute to friction in and smoothing of the global response to climate change—their efforts interact with more traditional ways of governing climate change in ways that can catalyze and facilitate changes in national and multilateral responses. Jane Gray revealed that their actions are always designed to spur national and international efforts at addressing climate change—subnational governments are test beds for taking aggressive actions.[36] The Climate Group creates friction and smoothing in a number of ways. First, there is direct interaction with the multilateral process. The Climate Group has been attending COPs since 2005, cohosting the 2005 Montreal Climate Leaders Summit and the 2009 Copenhagen Climate Leaders Summit, which showcased the ambitious actions being undertaken by Climate Group members. These activities at the COPs are not just show and tell presentations. The Climate Group has also actively advocated for greater recognition and incorporation of subnational actions in the multilateral negotiations. Thus at Copenhagen Climate Group members both signed a declaration outlining subnational commitments *and* (along with a number of other organizations) succeeded in getting subnational actors on the negotiating agenda.[37] These actions at the multilateral negotiations can serve as both an "irritant" that catalyzes movement at other levels of politics *and* a demonstration that the capacity to take climate action is growing.

Second, working within their network, The Climate Group works to find leverage points that will lead to a wider scope of action. Alison Hannon described an example of this process with her work in the U.S. Midwest. A strong international treaty needs U.S. support and implementation. In order for the United States to be at the forefront of this effort, there will need to be support from states in the Midwest, so The Climate Group supports state-level officials interested in progressive actions on climate change in an effort to leverage the U.S. position internationally.[38] This is a microcosm of their overarching "Breaking the Climate Deadlock" strategy that advocates for stringent global action by demonstrating the availability of solutions.[39] The claim is that

> sub-national leadership on climate change is not only generating much needed emissions reductions now, but is also having a valuable political

ripple effect. The success of policies trialed at this level gives confidence to national governments to emulate them.[40]

In addition at the Copenhagen Climate Leaders Summit, The Climate Group facilitated the signing of 12 agreements between regions in developed and developing countries to support climate initiatives in developing country regions.[41]

Finally, The Climate Group's transnational technology deployment activities are creating a need for new kinds of multilateral cooperation and coordination. Jessup recounts how they are running into some obstacles in India as they work to deploy the LED pilot programs for street lights. Funding for implementing the pilot project in Indian cities has come from the Indian central government and they were enthusiastic about wanting to do a large scale project, but they are loath to spend government funds on technology not produced in India—which is a problem because the best manufacturers of LED lighting (that would have the best chance of succeeding in a pilot program and thus increasing the chance of driving wider adoption) are not yet in India. Developing relationships from within The Climate Group's network and with the Indian government was key to making this work.[42]

Even so, Jessup sees the need to get international agencies focused on developing and coordinating these kinds of global trials so that there can be both local buy-in and a free flow of technology across borders.[43] A global trial of technology deployment creates the need for coordination at different scales—friction that can lead to broader cooperative efforts on climate change and smoothing that arises from demonstrating how technology can be deployed efficiently and effectively.

The Climate Registry: Building the Infrastructure for a Carbon Market

Similar to that of The Climate Group, the origin of the Climate Registry can be traced to a specific beginning, this time in legislation. The earliest incarnation of the initiative that would become the Climate Registry was the California Climate Action Registry, which was created—at the behest of industry leaders who "saw the eventual regulation of greenhouse gas emissions and wanted to protect their early actions to reduce emissions by having a credible and accurate record of their profiles and baselines"—by the California state legislature in 2001 through state bills 1771 and 527.[44] The California Climate Action Registry was originally designed to support California businesses and governments in measuring their carbon footprints and recording their efforts to reduce emissions. It was a voluntary reporting mechanism. Corporations and businesses interested in measuring emissions with an eye toward gaining credit for taking early action could opt to work with the California Climate Action Registry, but no one had to. The registry was going along smoothly in the early 2000s. Among its 23 charter members were large corporations like British Petroleum and Pacific Gas and Electric, as well as major municipalities like Los Angeles, Sacramento, and San Diego. Heavy corporate hitters like Kodak, Reliant Energy, Shell, Verizon, and Xerox subsequently joined the registry as well.[45] The registry was also a noted participant at the 2005 Montreal COP—celebrated as part of the progressive action California was pursuing in contrast to the recalcitrance of the U.S. federal government.[46]

But California wanted to do more than encourage voluntary reporting. The aforementioned negotiations over AB32—California's landmark climate legislation signed in 2006—included provisions for mandatory reporting, and the state senate leaders who were pushing AB32 wanted to scrap the California Climate Action Registry in order to eliminate the voluntary greenhouse gas emission reporting option.[47] So those involved in the California Climate Action Registry—its board of directors and staff—needed to make a choice between dissolving or evolving. They chose to evolve . . . and expand. They were aided in this decision by not only a compromise in the California legislature whereby the California Climate Action Registry would remain available for those not subject to mandatory reporting under AB32 but also the legislature's decision to accept emissions registered with the California Climate Action Registry as the baseline for the mandatory reporting on the horizon.[48]

Buoyed by momentum from the Montreal COP (California's delegation to the negotiations was larger than the U.S. federal delegation),[49] those working on the California Climate Action Registry decided to take the registry continental. Alex Carr, regional director in Canada, recalls that there was recognition that the California Climate Action Registry was a good model and that in 2006 people began thinking ahead to a continental scheme or at least interoperability of regional climate activities—a North American registry would thus be needed as part of the infrastructure for such activity. One novel approach—the California Climate Action Registry—seeded the ground for another novel approach. The Climate Registry was born, and with significant enthusiasm. Thirty-one U.S. states and two Canadian provinces were charter members of the Climate Registry.[50]

The Climate Registry and its parent the California Climate Action Registry follow chapter 3's script quite nicely. Urgency about climate change and frustration with the lack of federal action motivated the development of these registries. Of course, initiators (the state of California and then multiple U.S. states and Canadian provinces) and implementers (municipal governments and corporations) have slightly different takes on these issues. Corporations were concerned about receiving credit for early action on emissions reductions, and pushing for a registry filled a functional need. Corporations were/are convinced that market mechanisms to deal with climate change are on the horizon at some level and want to position themselves well for the coming regulations. Since measuring carbon is a prerequisite for doing anything in the carbon markets, especially to receive credit for early action, seeking the development of a registry was a natural course of action. Because there was a dearth of action at the federal level, it made sense to pursue a state-level registry.

Subnational authorities that responded (at least in part) to corporate desires were also similarly motivated by urgency and the lack of federal action, but there is the added motivation of achieving and protecting political authority. Steve Schiller, a member of the California Climate Action Registry's executive board and one of the founders of the Climate Registry, argues that there has been a natural progression from voluntary to regulated climate action.[51] The voluntary California Climate Action Registry gave way to mandated emissions reporting under AB32. The voluntary Climate Registry, in place across U.S. states and Canadian provinces, has given way to mandatory reporting for large emitters. Working on and with the Climate Registry has helped subnational governments to become familiar with greenhouse gas reporting and quantification issues, which helps them to more fully engage with the federal government on reporting. To be a member of the WCI you also have to be a member of the Climate Registry and Alex Carr observes, "We try to represent the interests of states and provinces to the federal government with respect to greenhouse gas reporting."[52]

In this sense, the Climate Registry serves multiple purposes. The first and most obvious is a functional role. It provides a way for corporations and subnational governments (cities, counties, states/provinces) to measure and record greenhouse gas emissions. They have developed a series of reporting protocols, laying out the methodology for measuring carbon footprints from both direct and indirect emissions.[53] In that they are a repository for third-party verified emissions inventories, implementing actors can also demonstrate reductions made over time. In this way, the Climate Registry is able to standardize the process of measuring greenhouse gas inventories for a range of actors (both those likely to be regulated in national or regional policies and those entities that will not be regulated) and recognize leaders on climate change, especially corporate leaders. In fact, the number 1 reason that implementing actors join the Climate Registry is to be recognized for climate leadership.[54] The Climate Registry is thus literally building the infrastructure for wider responses to climate change, especially through market measures. Robyn Camp, vice president of programs, notes how organizations and corporations become more educated on greenhouse gas emissions through participating in the registry. The Climate Registry is making it widely possible to measure and account for emissions—a necessary capacity for taking concerted efforts to reduce emissions and perhaps profit from such reductions.[55]

The second role is also about building infrastructure, but it is political infrastructure. Developing the Climate Registry has provided a powerful platform for subnational governments to engage with the federal governments in the United States and Canada as they (potentially) develop national responses to climate change. When the U.S. EPA was in the process of putting together its Mandatory Reporting of Greenhouse Gases Rule in 2009, the Climate Registry played a role in the process. During the public comment period, numerous corporations and subnational governments (in addition to staff from the Climate Registry itself) provided comments urging the EPA to adopt a number of the Climate Registry's procedures.[56] Interested parties urged the EPA to adopt the third party verification at the core of the Climate Registry's inventory protocols and to even allow entities participating in the Climate Registry to be exempt from the EPA's mandatory reporting requirements.[57] Proponents were not successful in getting all of the Climate Registry's procedures adopted, but the advocacy on its behalf from multiple sectors and types of actors demonstrates the kind of influence and relationships this experiment is developing.[58]

And even without wholesale adoption of Climate Registry mechanisms, it arguably influenced the EPA as it developed the reporting rule. In a response to a comment from the Massachusetts Department of Environmental Protection about reporting requirements, the EPA stated:

> EPA agrees that an already established reporting format should be used to submit the required data. To that end, EPA will be modifying the Consolidated Emissions Reporting Schema (CERS) to handle the reporting requirements of this rule. . . . CERS was jointly developed by EPA's Office of Air Quality Planning and Standards and Office of Atmospheric Protection, and The Climate Registry in conjunction with State, Local, and Tribal air pollution control agencies, and industry representatives.[59]

In addition, while the EPA ultimately did not waive reporting requirements for members of the Climate Registry, it did "consider waiving verification" for them and ultimately pledged to "develop coordinated verification approaches among programs."[60] This experiment is shifting the agenda for how the United States responds to climate change, and it is doing so by building infrastructure for the response.

The Climate Registry's interactions are not restricted to the U.S. federal government. Given that its board of directors consists of all state or provincial officials, it is natural that the Climate Registry works closely with regional emissions trading venues. It is working with the WCI to develop its regional reporting database.[61] Midwestern Greenhouse Gas Reduction Accord's advisory committee, which is developing the rules for the potential Midwestern cap and trade system, recommends using the Climate Registry protocols for calculating emissions and their reporting framework to support the Accord's mandatory emissions reporting requirements.[62] In fact, the Climate Registry is an experiment that is developing at the seams of voluntary and regulated activity. It has developed a "Common Framework for Mandatory Reporting" that allows subnational governments in the United States, Canada, and Mexico, along with Native American nations, to manage their mandatory reporting programs.[63] This is a key example of how experiments work together across governance models in a potential division of labor.

The work of the Climate Registry serves an explicit smoothing function in the broader global response to climate change, building a broad base of infrastructure that has the potential facilitate action at multiple levels. By laying out procedures and protocols for measuring carbon emissions at local and industrial plant levels, the Climate Registry literally makes it possible to formulate a number of climate actions. And it is not doing so in isolation. Robyn Camp notes that the Climate Registry is collaborating with ICLEI on municipal reporting protocols.[64] They are also connected with another kind of emissions reporting experiment, the Climate Disclosure Project that works to get large corporations to disclose their carbon exposure, through their joint membership in the Climate Disclosure Standards Board.[65] The Climate Registry has even participated in discussions about globalizing their model of greenhouse gas registry. They are currently working with a Chinese NGO and the environmental department in Israel to develop registries in those countries. Camp notes that the Climate Registry has no plans to become a global registry, but that they are participating in discussions about launching similar registries in other places. Infrastructure building is going global.[66]

C40: Taking Voluntary Municipal Action Transnational[67]

The organization that would become C40, the Large Cities Climate Leadership Group, was founded on the initiative of the mayor of London in October 2005 when the leaders of 18 major cities met in London to discuss ways cities could cooperate and support one another in pursuing action on climate change, swelling the ranks of city-led climate governance experiments like ICLEI's CCP and the U.S. Mayors' Conference Climate Protection Agreement. Drawing on a select group of 40 member cities and 17 associate cities from across the world, the C40 group partnered with the Clinton Climate Initiative in August 2006, securing additional resources with which to pursue their activities of capacity building, information sharing, and project implementation. C40 is one of the most visible voluntary action experiments, and its experience demonstrates how cities have come to see themselves as authoritative actors and how city networks have played a key role in the development of the experimental governance system.

The genesis of C40 is another interesting example of the novelty breeds uncertainty which breeds novelty dynamic discussed in chapter 3. By 2005, the idea of city networks designed to respond to climate change had already emerged. The CCP program of ICLEI had been running for over a decade, as had the Climate Alliance. Klimatkommunerna emerged in 2003 among Swedish cities, and the U.S. Mayors' Conference Climate Protection Agreement preceded the founding of C40 by eight months. The initiators of C40 drew explicitly on some of these experiences in designing the experiment. Phil Jessup, founding director of the CCP program, was seconded to the city of London at the time and worked closely with Deputy Mayor Nicky Gavron from London to bring together large cities—C8 was the original idea, which then expanded to C20 and eventually C40.[68] In some ways, C40 emerged not only in response to the issue of climate change and governance problems in the multilateral governance system but also in response to activities in the experimental system itself. The CCP program was not making inroads into larger cities, and while the networking, exchange of best practice, and action planning that it called for was considered an advance, there was also the feeling that it was too broad a program.[69] The experience and existence of one experiment in part generated the impetus for another.

The specific decision to innovate and form C40 was motivated by a combination of factors. Certainly there was a sense of urgency about climate change. The justification of the group's formation was framed in terms of the need to take action to tackle global warming because other actors were not taking strong or effective enough steps. Commenting on the specifics of Toronto's involvement, Mary MacDonald noted the growing tension between citizen interest in addressing climate change and national inaction in Canada. This made a transnational city network attractive: "You don't want to be a stand-alone all the time so you look to other cities."[70] Similar desire to motivate action at other levels by banding together has always permeated C40's outreach. The communiqué issued from the first meeting in October 2005 urged nation-states to work toward a new post–Kyoto Protocol treaty, while also asserting, "WE are ready to take action and join other cities, regions, states, provinces, national governments, and corporations around the world to lead the way."[71] The initial sense of urgency that drove the founding of C40 has increased as the multilateral process has floundered. The host mayor of Seoul, South Korea, exhorted his fellow mayors at the 2009 C40 summit with these words: "What makes us even more desperate is the fact that the result of our action will decide the future of humankind. As you may all agree, we are here today to turn that desperation into hope."[72]

Yet as with other experiments, a concern with climate change is not the only motivation to innovate or to join experiments. Cities are embedded in larger political structures within which they compete for resources and authority, and devolution dynamics whereby responsibilities for providing services have moved from national and state governments to municipalities have simultaneously strained cities' resources while also emboldening their claims to authority.[73] The C40 network thus also emerged to play a role in these politics as well, particularly given the initial leadership by Mayor Livingston in London, a controversial figure in the British political establishment, and his approach to seeking to carve out a distinctive political role for the new Greater London Authority.[74] This political nature of the C40 has become a defining feature of the C40 experiment. At the Seoul summit, David Miller (mayor of Toronto and current chair of C40) argued that "national governments must engage, empower, and resource, their cities so that together we can do more to combat climate change."[75]

"Engage, empower, and resource" has become the C40 mantra, repeated again at the 2010 Climate Summit for Mayors in Copenhagen. The communiqué that emerged at the end of the summit articulated the continuing role of complex motivations that drive this network of major cities to innovate—a blend of sincere concern about addressing climate change with goals of enhancing the position and authority of cities:

> Over the years, we have undertaken climate strategies that were often more ambitious than national action. We intend to continue doing so but require stronger cooperation between national and local governments. A cooperation that promotes the involvement of cities in reaching our common goal: a global low-carbon, climate resilient future.
>
> We are prepared to collaborate, innovate, and try even harder. Our message to national governments is simple: agree on ambitious targets and start reducing now—and be confident that if cities are engaged, empowered, and given the right resources we will deliver on our commitments.[76]

C40 does not directly regulate participating cities—it is a quintessential voluntary action experiment. C40 relies on cities being self-motivated and informal accountability in the community of mayors.[77] Even so, C40 has established a network that works to enhance its members' abilities to pursue city-level action. Like many other voluntary action experiments, C40 engages in networking and planning core functions. Two aspects of C40 functions make it somewhat unique among city networks and voluntary action experiments: its explicit association with another voluntary action experiment (the Clinton Climate Initiative) and the prominence of the city members. The Clinton Climate Initiative is the implementing actor for C40, which means that it facilitates carrying out the voluntary actions agreed to by C40 cities. This is occurring in a number of dimensions (some of which are explored in greater detail in chapter 5):

> Pooling the buying power of cities. This will help lower the prices of energy saving products and hasten development and uptake of new

energy saving technologies. The consortium will partner with vendors leading to lower production and delivery costs and therefore lower prices. Key product categories will include building materials, systems, and controls; traffic and street lighting; clean buses and waste disposal trucks; and waste-to-energy systems.

Mobilising expert assistance to help cities develop and implement programmes that will lead to reduced energy use and lower greenhouse gas emissions. Technical help will be provided in areas including building efficiency, cleaner transport, renewable energy production, waste management, and water and sanitation systems.[78]

All of these functions are geared toward enhancing member cities' climate action plans and fostering best practices in activities that cities can directly control: building codes and retrofitting standards; energy efficiency and clean energy in municipal buildings and transportation fleets, local clean energy generation, urban planning for public transportation and congestion, waste management, and water distribution.[79] In addition, C40 is expanding its activities by partnering with the new Carbon Finance Capacity Building program (another voluntary action experiment). C40 is looking bring cities into the growing world of carbon markets.[80]

But like The Climate Group, C40 is also building a powerful community of implementing actors. Mary MacDonald observes: "there's a real camaraderie among mayors," and Rohit Aggarwala from the New York City Mayor's Office explains that the knowledge networks and personal relationships developed through C40 have been important in supporting New York's climate change policies.[81] Much like The Climate Group's high-publicity programs, Aggarwala also recognizes the "marquee value" of initiatives supported by C40 and the Clinton Climate Initiative.[82] The high profile of C40 means that it has an advantage in creating the friction and smoothing discussed in chapter 3—both through its advocacy efforts and actions.

C40 is also active in the UN negotiations, sending delegations that seek to influence the development of the multilateral regime. They continually ask national governments not only to include cities in the global cooperative process (engage) but also to recognize that "Cities Act" and empower and resource them to move forward with climate policy. But C40 has much higher ambitions than merely seeking to lobby states in the midst of their negotiations. On the contrary, C40 hopes to lead, with Miller claiming that "if governments talk about reducing CO2, cities are the ones that show how it's done."[83] C40 is claiming to act in the role of laboratory of democracy that scholars have identified for subnational governments.[84] This is an explicitly stated goal on their website, where C40 argues that "cities are often centres of new thinking and policy innovation—cities are in a great position to lead the way for others to follow."[85] While the original wave of municipal experiments was significantly concerned with taking action in the face of national recalcitrance and multilateral stalemate, this has evolved into a call for cities to lead no matter what the multilateral process produces and for recognition that "the battle to prevent catastrophic climate change will be won or lost in our cities."[86] Innovation serves as an irritant in the system—that is, other political actors are forced to deal with the ramifications of innovations that go beyond the boundaries of C40 cities—but it also provides the means of soothing irritation, as the programs that C40 initiates are intended to make implementing the global response to climate change more feasible.

C40's actions and interactions are also building the experimental system through both cooperation and competition, with resulting redundancy. Cooperation across experiments brought about C40 in important ways—the experience of the CCP was crucial, and The Climate Group worked on the original summit that brought the C20 into existence. Very quickly, C40 joined with the experimental Clinton Climate Initiative to carry out its plans and activities.[87] So networking among experiments is a key part of C40 existence (note that the 2009 leadership summit at Copenhagen was cosponsored by ICLEI). Yet there is competition as well. This was illustrated by the fact that C40's leadership summit at Copenhagen occurred at the same time as the leadership summit sponsored by The Climate Group. In this sense, competition is breeding redundancy in the experimental system, which many consider to be a good thing. The Climate Group and C40/Clinton Climate Initiative are engaged in similar programs to deploy technology in city networks (discussed in chapter 5) and at Copenhagen, Mayor Miller discussed the need for cities to partner with business, arguing that "cities cannot fight climate change on their own. We need all kinds of partnerships to serve the public interest. As Chair of the C40, I must emphasize that in order to fight climate change, we need a global partnership involving national governments, the public and businesses."[88] This sounds a great deal like the network that The Climate Group has been developing since 2004.

Chicago Climate Exchange: Making Accountable Emissions Trading Work

Akin to C40, it is impossible to discuss the origins of the CCX without referring to innovation in other places and times. In this case, however, the inspiration was not drawn from an existing climate governance experiment but from the innovative idea of emissions trading itself—developing cap and trade systems for pursuing efficient, low-cost environmental protection, especially for air pollution. In theory, cap and trade mechanisms drive down the cost of pollution abatement because those that can efficiently reduce pollution will go below their permitted levels and sell remaining permits to members that cannot meet their allowances (this works when the traded permits are less expensive than the abatement measures that would be undertaken by the less efficient members). While this mechanism has engendered significant debate among economists and policy-makers, some of whom favor carbon taxes instead of cap and trade measures, it does have a number of potentially positive qualities for application to climate change.[89] First, if well enforced, the declining cap guarantees that emissions are reduced. Second, in tying the cost of permits to those encountered by the most efficient reducers, it drives down the cost of attaining reductions. Finally, it has the potential to spur investment in climate-friendly technologies both to attain compliance with the cap and because the authority issuing the permits can sell or auction them (ideally they would, according to most economists) and use the revenue to invest in climate friendly technology.[90]

Cap and trade mechanisms were originally put into practice in the 1980s to address the problem of acid rain in North America and Europe,[91] but they were not introduced into discussions of climate change until the Kyoto Protocol negotiations in the mid-1990s (though earlier discussions in the OECD focused on cap and trade mechanisms along with other market measures).[92] At the insistence of the United States and many business interests—and over the objections of the European Union and many developing countries and environmentalists— this "flexible mechanism" was included in the Kyoto Protocol as one tool for states to achieve their emissions reduction commitments.[93] However, the Kyoto Protocol and subsequent Marrakech Accords provided little guidance on how emissions trading should take place. The Kyoto Protocol catalyzed the development of cap and trade as a governance mechanism for climate change, but it did little more than open space for developing this mechanism. In the intervening decade, actors at multiple levels (cities, states and provinces, corporations, and individual nation-states) have taken up the challenge of designing and implementing cap and trade systems.[94]

It was in this context, where market mechanisms and especially cap and trade were becoming inextricably linked to action on climate change, that discussions arose over the development of the CCX. Officially, the design of the CCX experiment began in 2000,[95] but the development of the CCX has a deeper history, and its intellectual roots go back to the early days of the climate negotiations in the early 1990s and efforts to combat air pollution in the United States. Already in 1992 Richard Sandor, the economist from Northwestern University who was a principal designer of CCX, contributed to a UN volume on global emissions trading, *Combating Global Warming: Study on a Global System of Tradeable Carbon Emission Entitlements*.[96] Meanwhile, Michael Walsh, the other key founder of CCX, was working with the Chicago Board of Trade designing and administering sulfur dioxide auctions for acid rain cap and trade programs in North America.

As the UN negotiations ramped up in the mid-1990s, so did the profile of cap and trade as a favored market mechanism for climate change. Michael Walsh recalls how he started sensing interest among private and nongovernmental sectors to "get going" on emissions trading—testing things out and learning how to do cap and trade.[97] Even before the Kyoto Protocol was signed, the UN Conference on Trade and Development was organizing workshops on emissions trading, a "Greenhouse Gas Emissions Trading Forum" with the first one hosted by Sandor and Walsh in Chicago in June 1997. The goal was to both advocate for market mechanisms in the climate negotiations and familiarize actors with this tool—so as to "provide timely institutional support to interested governments, corporations and non-governmental organizations, for the development and implementation of the initial-phase of an international greenhouse gas emissions market."[98]

The motivation for moving forward with an experimental cap and trade system is thus relatively clear and matches the script evident in other experiments. The negotiation and signing of the Kyoto Protocol raised the profile of climate change and seemed to promise action to come. Most of the early discussions surrounded a potential global trading system that was to be part of the Kyoto Protocol. This moved actors (initiating and implementing) to begin responding to climate change in anticipation of or in preparation for global and and/or national action. When international and North American *in*action became increasingly obvious in the late 1990s, initiatives became more independent from traditional governance practices and morphed into climate governance experiments, but with an eye toward driving broader policy if and when it does develop at the national and global levels.

In 2000, the Joyce Foundation provided funding for Sandor and Walsh to study the feasibility of a North American trading system. They began to attend environmental conferences in order to find out "who wants to be serious" about implementing cap and trade.[99] They did not want to argue about the science of climate change or debate the merits of a market based approach. They were instead "only going to argue about how" to get cap and trade up and running.[100] With continued funding from the Joyce Foundation in 2001–2002, Walsh and Sandor along with "more than one hundred professionals in the corporate, public, non-governmental and academic sectors" developed the set of rules "that would underpin and shape a pilot reduction and trading design."[101] In 2003, CCX was launched, with legally binding reduction commitments and trading among 13 charter members, including subnational governments (city of Chicago) and major corporate actors (Motorola, Ford, DuPont).

The CCX was built essentially from scratch, as it had no government mandate and very few existing trading system examples to draw on. In 2001 when the CCX design process got under way, Shell and BP were running internal corporate cap and trade systems, and Denmark had a national cap and trade system.[102]

What the CCX designers set out to do was to learn by doing and to move quickly to get a workable set of rules rather than a comprehensive set of rules at the outset. As Walsh recalls,

> a number of manufacturers, agricultural folks, forest product folks, etc., all had their own set of issues so we had them break out into working groups [topical groups]. We handed out a Chicago Accord—a 15-page summary of the core rules, and asked people to sign up. When they signed up, they understood that the Accord was just a summary and that a detailed rule book would be formed (2002), which would articulate further the details in the book. They formed a governance committee for each topical area and kicked off the auction.[103]

The results of these early discussions and the fine-tuning that came later produced a comprehensive cap and trade system that is voluntary for entities to join but legally binding once they sign a contract.[104] It covers all six greenhouse gases and has a legally binding target. Because CCX is a voluntary cap and trade system, the allocation system is a bit different from the regulated cap and trade systems like the EU emissions trading system or the RGGI in the U.S. Northeast. The allocation is more individualized and is calculated by examining each entity's baseline emissions (the average of an entity's emissions from 1998 to 2001) and its reduction target (1% below baseline per year, down to 6% below baseline target for 2010). So each year, members are given their allowances (baseline emissions minus the target percent) and must in return turn in permits (called carbon financial instruments in CCX) annually that match their target. If their actual emissions, which are verified and audited by a third party,[105] exceed their target, they need to purchase additional permits from other members through the CCX.

The CCX is an accountable action experiment because while it is voluntary to join CCX, the contract that corporations and subnational governments sign when they join is legally binding. So compliance measures are crucial, but it turns out to not be much of an issue, according to Walsh. There have been some small problems, but the big compliance issue, when "somebody that does not achieve the emissions cuts internally doesn't want to buy credits," has yet to really materialize. While CCX has "had to put some pressure on . . . most folks who participate do so for reputational advancement" and comply.[106] This is evident in CCX's audit reports, which show that members have indeed reached their reduction goals—CCX is one of the few climate governance experiments that can unequivocally claim to have demonstrated emissions reductions. According to the recently published 2008 compliance report, CCX members' emissions were collectively almost 11% below their stated objectives (1% per year reduction culminating in a 6% total reduction by 2010). Given that the CCX members' emissions equal those of the industrial emissions of Germany, this is no mean accomplishment.

This is positive in and of itself, but there is a broader purpose to CCX as well. From the outset CCX was designed explicitly as an experiment in cap and trade—working out the kinks to show how it could be done:

> it is widely observed that the best way to advance the process is to promptly begin trading, even if on a limited scale, so that institutions and skills can be built on the basis of real-world experience. Pilot greenhouse gas markets can offer a means to do just that.[107]

The long-term view was always on broader policy and adoption of cap and trade. As Walsh recalls, "we were trying to formulate a structure, and standards and rules that had a decent prospect of being in alignment with what governments would ultimately adopt."[108] The CCX is about pushing the boundaries and developing human capital for making emissions trading work—smoothing the way for national responses while creating friction, in that participation in CCX builds a constituency for national responses. It is an example of one of the conjectures discussed in chapter 3—that the experimental system could end up driving the agenda for the response to climate change at the national and global levels. In opining about the future of the CCX and the possibility of regulated cap and trade markets emerging in North America, Walsh noted:

> there a number of forks in the road. A number of the entities who are working with us will not become regulated entities, but who like what they are doing here with CCX, can show off their progress and hitting their goals. In addition we always have a number of market initiatives under way which are far from being legislated into being-helping to shape policy, but the policy is many years out. This is where a lot of the fun is. The U.S. government only gets to doing things when they have already been done by states, cities or private actors. Washington only does things after enough acceptance has been built up.[109]

Whether this acceptance will continue to build in the near term is an open question. The political climate in 2010 was a challenging one for carbon markets in North America, and it may remain so for the near term. In fact, CCX will cease its trading functions in North America at the end of its second Phase in late 2010, continuing its existence as an offsets registry.[110] This does not negate the important role that the CCX has played in the experimental system of governance—pioneering private emissions trading and thereby building the capacity of numerous corporations and other actors to engage in carbon markets. However, it is a stark reminder that experiments evolve and sometimes fail (though it is too early to judge whether the cessation of trading is a "failure"). There is a quality of "trying something out" in the experimental governance system, and consequently the existence and functioning of some individual experiments may be fleeting. Yet even so, the impact of an experiment may persist and expand, individually and collectively as part of the larger processes of experimentation and self-organization, even after it ceases to function or evolves into something different.

Conclusion

When the details of 4 experiments are examined closely, we can see that the account developed in chapter 3 is skeletal but compelling. The range of personalities, diverse motivations, and idiosyncratic details embedded in just these 4 experiments are fascinating and demonstrate just how complex the world of climate governance experimentation has become. Yet the theoretical account seems relevant even at this more microscopic level, as the details of these 4 experiments broadly fit with the expectations found in Chapter 3.

Across the board, these experiments reacted to the negotiation and signing of the Kyoto Protocol or stalemate in the multilateral process, sometimes both. It is fairly safe to say that perhaps none of these experiments would exist had the Kyoto Protocol developed into a stable, effective driver of global climate governance. Both The Climate Group and C40 emerged after it was clear that the Kyoto Protocol process was stalemated, and they appealed to implementing actors who wanted to move forward on climate change (for a variety of reasons). The Climate Registry and the CCX saw their background motivations change from responding to the possibilities of a broader global response to climate change to acting in lieu of a global or national response. The conditions posited to generate uncertainty in the multilateral governance system arguably played a role in motivating the emergence of these four experiments.

There is also evidence that a nascent experimental system is being built through experiments' interactions and relationships (both competitive and cooperative). Experiments are taking on defined roles and providing (unconsciously) niches for one another in the experimental system. The Climate Group helped to found C40, which emerged in part in reaction to the kind of activities the CCP program both was and was not undertaking. Then, after allying with the Clinton Climate Initiative, C40 has embarked on programs that are similar and in some cases competitive with The Climate Group. The Climate Registry has emerged on a North American scale, providing registry methodologies and services developed at least partly in conjunction with the Voluntary Carbon Standard, an effort of The Climate Group and the World Business Council on

Sustainable Development. The Climate Registry's services are being considered for uptake in the experimental regional cap and trade systems in North America that may be competitive with the CCX (at least they will be competing in terms of the policy-making at the federal level). And this is just a brief recitation of connections among 4—admittedly prominent—experiments. It is plausible that an experimental system of governance, bubbling away with only tenuous connections to global and national climate change responses, is indeed emerging.

Finally, these vignettes provide the beginnings of insight on the conjectures that closed chapter 3. It is still too early to tell definitively whether a division of labor or Tiebout sorting is emerging in the experimental system, but interactions within and across the governance models do indicate that a division of labor is a plausible outcome. Because experiments are filling different roles in the experimental system, a number of natural connections and synergies are emerging. We can also begin to observe how experiments and the experimental system are interacting and will continue to interact with the traditional governance of climate change that is centered on nation-states and treaty-making. Experiments are generating friction. The Climate Group, C40, and the CCX are pushing boundaries, going well beyond what nation-states and the multilateral process are ready to countenance in the response to climate change. If they are able to scale up their operations and effectively pursue the strategies they have put forward, the potential is there to catalyze broader action. Should broader action materialize, it will go more smoothly because of experimental activities. The Climate Registry is providing infrastructure not only for other experiments but also for national and international governance. C40 trumpets its readiness to implement global treaties and national plans if only nation-states would come to agreement. The CCX explicitly sees itself in an experimental role—working through cap and trade mechanisms and providing the experience that can be drawn on in creating federal and even global emissions trading.

But are these and other climate governance experiments effective? This chapter could be seen as building a positive case not so much for the plausibility of the account developed in chapter 3 as for the experimental governance system itself. I consciously presented the experiments in a straightforward way that emphasized their own stated potentials for catalyzing action, rather than in an overtly critical manner, because the point here is to assess what is possible in the experimental system. Yet it is impossible to evade questions of effectiveness. Obviously, assessing the effectiveness of climate governance experimentation depends on the rubric one uses to assess effectiveness.[111] There is simply very little reliable data to assess their effectiveness on the most important criteria—reduction of greenhouse gases. Besides the CCX, it is not obvious what kind of accomplishments the other 3 have in terms of emissions reductions. Further, direct emissions reductions are not really the specific goals of the other 3.

A different measure of effectiveness would assess these experiments in terms of their more proximate goals—their ability to diffuse information and build communities of like-minded actors who take actions on climate change. Measuring these criteria is a vexing problem. Is C40 effective because it is no longer C20? Is the Climate Registry effective because 41 U.S. states and all Canadian provinces and territories are members? Is The Climate Group effective because it can pilot programs in a range contexts and bring to bear its network for both governmental and corporate initiatives? One goal of this book is to lay out the structure of climate governance experimentation and start thinking about the ways an effective response to climate change *could* emerge collectively from what are new and relatively small-scale experimental activities—how an experimental climate governance system can provide the friction necessary to generate change in the global response to climate change.

The vignettes presented here provide a sense of the possibilities and justification for further examining the development of the experimental system of governance. They demonstrate how individual experiments emerge and how links and relationships in the experimental world and between the experimental and multilateral governance systems are beginning to be forged. The next two chapters go further and explore the relationships and structure developing around the deployment of technology in cities (chapter 5) and in carbon markets (chapter 6)—sites of experimental governance where multiple experiments are converging and altering what counts as the global response to climate change.

5

Experimenting with Cities and Technology Deployment

> Over the last decade, academic research has helped to consolidate our understanding of the role of cities in addressing climate change. In both the scholarly debate and in practice, cities have been recognized as an important site of climate governance.
> —Michele Betsill and Harriet Bulkeley

It is no longer novel to examine municipal responses to climate change. Cities and climate change have become a cottage academic industry.[1] This is not a slight—some of the best work on the politics of climate change focuses on cities in order to showcase the multiple scales at which the global response to climate change is developing and how an issue can simultaneously exist in both local (individual cities) and global (transnational city networks) contexts. Collectively, this work has closely tracked the evolution of cities on the global climate change stage from an afterthought to central player in the aftermath of the Copenhagen negotiations in 2009.[2] City actions were some of the few bright spots in Copenhagen, and with their mantra of "Cities Act" and vows to move forward with or without a global treaty, municipal activities seem destined to continue to play a key role in the global response.

This chapter will not rehearse what is already a full analytic agenda concerned with city-level responses to climate change, but it will briefly look at the lessons learned from city-oriented studies. Specifically, it is important to understand in general why cities have gotten into the business of climate change, why they form networks, and what relevance they have for the broader response to climate change. I then examine a particular cluster of activities that attempt to deploy climate-friendly technology in city networks. Of the 58 experiments, 14 are at least partially engaged in technology deployment in cities.

Some experiments are cooperating, others are competing, and still others are working in parallel with minimal interaction. The dynamics occurring in this cluster are just what we would expect from a self-organizing system that is transforming novelty into normalcy. However, the critics of city-led climate responses are unlikely to find much to shake their skepticism about the effectiveness of municipal responses to climate change in the descriptions that follow. Scaling up municipal initiatives and producing significant emissions reductions through them, especially in the absence of significant national and/or global action, will remain a daunting task. City/technology initiatives may need to rely on the indirect effects of friction and smoothing to have a significant impact on the global response to climate change.

Cities and Climate Change: Strange Bedfellows?

Diverse quantitative indicators imply that cities have a large and growing impact on global greenhouse gas emissions:

- "Home to half the world's population and growing rapidly, cities consume over two-thirds of the world's energy and account for more than 70 percent of global CO2 emissions."[3]
- "Recognizing that at present over 50% of the world's population lives in cities, which now account for 75% of global energy consumption and 80% of global greenhouse gas emissions and at this rate, by 2030, two-thirds of the world's population is predicted to live in urban areas."[4]
- "By 2030, two thirds of humanity will live in urban centers, where today more than 50% of the world's population lives and more than 75% of all energy is consumed. All cities are highly vulnerable to the impacts of climate change, especially fast growing cities in developing countries."[5]
- "By 2030, three-quarters of the world's population will be urban. . . . Cities generate close to 80 percent of all carbon dioxide and significant amounts of other greenhouse gases mainly through energy generation, vehicles, industry and biomass use."[6]

Even so, many studies of cities and climate change begin by asking versions of the question "Why on earth would cities get involved in the response to climate change?" They ask because on the face of it there seems to be little incentive or capacity for cities to address what is perhaps the only truly global problem that humanity faces. The initial skepticism arises from economic analysis—actions undertaken by individual cities will engender private costs (i.e., the individual city will have to pay some cost to enact climate policies) but only provide negligible and diffuse climate benefits. Given their relatively limited capacity and scope for action, the climate-related benefit from any particular city or group of cities doing anything is small and shared by the whole globe, while the costs of doing something are born only by the citizens and cities that undertaken action. From this perspective, municipal action on climate change is irrational.

Yet cities do act on climate change, individually and in concert, in increasingly large numbers. The literature on cities and climate change has managed to make sense of this action in a number of ways—city action is not as irrational as it first appears. A number of studies have assailed the assumptions on which the skepticism is based. It turns out that cities' actions on climate change entail, in many cases, negative costs (actually saving money for cities) and cobenefits in terms of health and economic development.[7] Rohit Aggarwala formerly of the Mayor's office in New York City recalled that New York's drive on climate began, in part, over concerns about air quality.[8] This means that the benefit from addressing climate change not only is shared globally, but that there are indirect, local benefits as well—all of which go significantly beyond the benefits of emissions reductions themselves. Molly Webb, head of strategic engagement at The Climate Group, stresses that even in the post-Copenhagen doldrums, cities are moving forward on climate because they have nonclimate reasons to pursue energy efficiency and other climate related policies.[9] In addition, 41% of respondents in a U.S. Mayors' Conference survey claim that the "primary focus of their city's climate protection strategy . . . is making governmental operations and services more energy efficient."[10]

Others point to the symbolism and strategy behind city initiatives on climate change. Some cities see value in being leaders on climate change and in undertaking actions that will create pressure for broader responses to climate change.[11] For instance, the C40 group of cities has dubbed itself a "Climate Leadership Group," and it works to push forward on climate action in cities and in the larger political jurisdictions in which cities are embedded. Richard Stewart argues that it is precisely these indirect effects that are motivating cities to act on climate change. City action is about forcing "the pace and stringency of national action."[12] This progressiveness on climate at the municipal level is pursued and made possible because cities are more accessible and responsive to pressure from civil society and citizen demands.[13]

The decision to form networks is another facet of the cities and climate change literature relevant to this discussion, though it has received less scholarly attention than the fact of municipal action itself. Scholars have begun to pay serious attention to the transnational municipal networks that exist between local grassroots efforts and national/global responses to climate change.[14] Consensus has yet to be reached on why cities form networks, though most studies point to instrumental concerns like the role of network learning, access to financial and political resources, improving the governing performance of individual cities in the networks, or enhancing the visibility and role of cities in national and multilateral processes.[15] At the recent Copenhagen negotiations, a C40 communiqué boldly declared:

> We, the mayors and governors of the world's leading cities, have joined together in Copenhagen in December 2009, at the Copenhagen Climate Summit for Mayors, to send a strong and united message to national governments: seal the deal in Copenhagen and acknowledge internationally the pivotal role of cities in fighting climate change.[16]

Harriet Bulkeley and Kristine Kern make a persuasive case that municipal networks are seen to accrue benefits to city members in terms of access to funding and knowledge, but caution that "they appear to be primarily networks of pioneers for pioneers."[17] Thus, the broader relevance of city networks is still an open question, and evaluating their effects is complicated by the fact that municipal networks "undertake a variety of governing activities" aimed at both enhancing city operations and influencing the external world.[18] Cities and their networks make strident arguments about their relevance in the global response, and scholars claim that "cities, rather than nation-states, may be the most appropriate arena in which to pursue policies to address" climate change.[19] Yet to this point in history and in the global response to climate change, while cities are certainly crucial players in the *generation* of climatic change, it is less immediately obvious how relevant cities and their networks are or have been in *addressing* climate change. While information exchange and networking are touted as positive steps, Kern and Bulkeley warn that these activities "are appealing in that they demand fewer network resources and less intervention, but this is countered by less certainty at the network level as to what is being achieved and for whom."[20]

In fact, most observers do not judge the relevance of city initiatives and networks on the amount of greenhouse gas emissions reduced.[21] It is difficult to obtain exact numbers on how much cities and city networks have reduced their emissions. Emissions calculators tailored to municipal needs have only recently become widely available. The U.S. Mayors' Conference reported that only 36% of member cities who completed their 2008 survey had completed an inventory of greenhouse gas emissions for city operations and only 25% completed an inventory of municipal emissions writ large.[22] Further, given the voluntary nature of the city network experiments examined here, there is little in the way of monitoring or enforcement that would mandate such reporting. Megan Meaney noted that the Canadian Partnership for Climate Protection has only recently begun asking their network members to report emissions profiles and reductions to ICLEI.[23] Their recent survey was aimed to "offer what has been lacking in previous reporting on climate mitigation efforts by local governments within Canada, namely quantifiable results of municipal and community greenhouse gas reduction measures." And in fact, they have some heartening news from the self-reporting of the member cities, finding that "to date, the cumulative annual savings reported by local governments since the database began is 1.4 million tonnes of GHGs; this is equal to removing approximately 325,600 light vehicles from the road."[24] Further:

> we know we have captured only a fraction of the measures being imple-
> mented by Partnership for Climate Protection participants. At present
> date there are 183 members in the PCP program, all of which are taking
> action on climate change mitigation. Yet, sixteen municipalities
> reported measures in the 2009 reporting period. To bridge that gap we
> must focus on building the capacity of the remaining 167 municipal-
> ities to track and report their GHG emission reduction activities.[25]

Emissions reductions may be the most discussed marker of climate action, but it is not the most visible, and it might not ultimately be the most important in the near term. The ultimate goal of climate action is redirection of the economy and society onto a low-carbon pathway. In this sense, indirect catalytic effects on the economy and political response to climate change may be better indica-tors of relevance.[26] In fact, a number of observers point directly to this, and it is evident that cities and their networks see this as important as well.[27] Putting pressure on other levels of government and stimulating technological develop-ment are oft-cited justifications for studying and pursuing municipal action on climate change.[28] Transforming the market and altering the relationship between carbon emissions and economic activity is the goal.[29] Cities and their networks do not necessarily get in the business of climate change because they believe that they can significantly reduce global greenhouse gas emissions on their own, but they do believe that they can play a catalytic role on a number of fronts that will move the global response to climate change forward:

> Even if local communities can't solve the global warming problem on
> their own, they are gathering important experience with policy op-
> tions, while sending an ever-stronger signal to state, federal, and inter-
> national policy-makers that action is possible, cost-effective, and
> politically supported.[30]

Technology Deployment Cluster

Cities and their networks have received academic and media attention with good reason. Whatever the ultimate efficacy of municipal action turns out to be, this is certainly an area of considerable momentum on climate change action. It is striking just how many experiments are engaged in some aspect of deploying technology in cities and city networks. Experiments from each of the four experimental models participate in this activity cluster. The range of technologies being developed and employed is impressive. Most experiments engaged in this area are linked to one or more additional climate governance experiments, either other experiments working in this activity cluster or experiments working in other clusters. Table 5.1 provides a brief synopsis of the experimental activity in this area.[31]

Three types of experimentation are evident in this activity cluster. First, there are the actual city networks—experiments that exist to link municipalities and facilitate their climate activities across jurisdictions. All of the major municipal networks (C40, Climate Alliance, Covenant of Mayors, EUROCITIES, ICLEI's CCP, and the U.S. Mayors' Conference Climate Protection Agreement) are represented in this activity cluster, and all are advocating and/or facilitating the deployment of various technologies in their networks. Second, there are a range of technology focused experiments (Climate Neutral Network and Connected Urban Development Program) that find municipalities and their networks a fruitful level of political organization in which to deploy their work. Finally, a number of experiments that seek to bring cities and corporations together work in this area (2degrees, Carbon Finance Capacity Building Partnership, Clinton Climate Initiative, The Climate Group, World Business Council for Sustainable Development).

These diverse experimental activities are held together by a commitment to deploying climate friendly technology in cities. There is a clump of experimental activity in this area that signals that self-organization in the experimental system is proceeding. Relationships among experiments are emerging around this activity. Multiple experiments with diverse approaches to deploying technology create redundancy. Deployment of technology has become a site of concentrated experimental activity and this gives structure to the experimental system beyond, and building on, the common liberal environmental foundation and functional differentiation evident in chapters 2 and 4. But this needs some unpacking. We need to understand both the practicalities of this activity—what technologies are being planned for or deployed in cities?—and the goals—to what end is the activity developing? With this background in place, we can then turn to the interactions and relationships developing in the experimental system.

Four broad areas of technology development and deployment consistently emerge in experimental activities:

- *Renewable Energy:* Renewable energy—solar, wind, biomass, biodiesel, geothermal, fuel cell—is a constant in discussions of climate policy. With the ultimate goal of decarbonizing the economy and energy sector, the need to

Table 5.1 **Experiments in the Municipal Technology Cluster**

Experiment	Governance Model	City/Technology Activities	Interactions with other Experiments*
2 Degrees	Networker	Network working groups on: Low Carbon Data Centers, Smart Cities, Low Carbon Real Estate	Cisco (initiator of Connected Urban Development Program)
C40	Voluntary Actor	A range of programs for building retrofits, lighting, mobility, Information and Communications Technology (ICT)	Connected Urban Development Program, Clinton Climate Initiative
Carbon Finance Capacity Building Programme	Voluntary Actor	Advice on generating offset credits from energy efficiency programs in cities	Supported by C40
Climate Alliance	Voluntary Actor	Renewable energy and transportation programs	US Conference of Mayors
Climate Neutral Network	Infrastructure Builder	ICT and energy efficiency in buildings	Clinton Climate Initiative, Cities for Climate Protection, WBCSD
Clinton Climate Initiative	Voluntary Actor	A range of programs for building retrofits, lighting, mobility, ICT	C40, Climate Neutral Network
Connected Urban Development Program	Voluntary Actor	ICT, Mobility, Energy Efficiency in buildings	The Climate Group, Clinton Climate Initiative, C40

continued

Table 5.1 (continued)

Experiment	Governance Model	City/Technology Activities	Interactions with other Experiments*
Covenant of Mayors	Accountable Actor	Sustainable Energy Plans	Climate Alliance, Klimatkommunerna
Eurocities Declaration	Voluntary Actor	ICT, Mobility	Cities for Climate Protection, Union of Baltic Cities, US Mayors Agreement on Climate Change
ICLEI—Cities for Climate Protection	Infrastructure Builder	A range of energy efficiency programs	Climate Neutral Network, C40, US Mayors Agreement of Climate Change
REEEP	Networker	Renewable Energy	
The Climate Group	Networker	SMART 2020	Connected Urban Development Program
US Mayors Climate Protection Agreement	Voluntary Actor	Best practice recognition for transportation, energy efficiency, building, and renewable energy programs.	Eurocities, Climate Alliance, ICLEI Cities for Climate Protection
World Business Council for Sustainable Development	Networker	Energy Efficiency in Buildings	Climate Neutral Net

* Note—these are just the formal experiment-experiment interactions listed by the experiments themselves. At another level—the way that individual cities interact with different experiments—the interactions are even denser. For instance, in a report on city leadership by The Climate Group, a blurb on Seattle notes its dealings with Climate Wise (The Climate Group 2005). Similarly, a celebration of city actions published by the US Mayors Conference, notes in a number of places the role of ICLEI's Cities for Climate Protection programs. The varied types of interactions will be discussed in greater detail below.

utilize and scale up renewable energies currently available and develop and deploy new ones is high on the priority lists. Cities are an important part of this discussion and trend. Two types of renewable energy projects are iconic in municipal climate governance experiments. There are a number of experiments encouraging the use of renewable energy in city operations and at the household/business level. Solar projects like the one in Boston highlighted by the US Mayors' Conference encourage the deployment of solar panels throughout the city.[32] Other projects use the purchasing power or control over city utilities to encourage renewable energy more broadly. Houston, for instance, has a plan in place to purchase 50% of its power from renewable sources such as wind power.[33]

- *Energy Efficiency/Building Retrofitting:* Energy efficiency, especially in buildings, is another crucial area of technological concern. The World Business Council on Sustainable Development estimates that 40% of global emissions arise from buildings.[34] This is a significant area of interest for cities for a number of reasons. First, given that cities own significant property and systems such as street lighting, energy efficiency programs have the potential to produce cobenefits regardless of climate outcomes. Second, cities that control building codes and can channel grant programs from national governments can implement green building requirements. Finally, energy efficiency drives in cities are a way to publicize climate action among citizens and businesses in a way that stresses the compatibility of economic growth and climate action—the path many cities have chosen.

- *Mobility:* Public transportation is another classic response to climate change at the municipal level. However, efforts to deploy climate-friendly technology in cities go beyond building more light rail or subway systems like the one for which Denver was given an award by the US Mayors' Conference.[35] Mobility has also come to be about city planning to make cities more walkable and bikeable as well as using technology to enhance the flow of traffic and make public transportation more efficient.

- *Information and Communications Technology (ICT):* This fourth area of technological development and deployment may be the least well known, but it is growing rapidly. At one level it is akin to energy efficiency, in that cities own a lot of ICT infrastructure and there are projects that seek to get cities to understand the footprint of these technologies and use them more efficiently. More ambitious programs look to ICT as a way to coordinate activities in the other three areas—deploying ICT to use power in smarter ways, enhance efficiency measures, and design new transportation systems. I will explore this kind of activity in greater detail below.

Again, addressing climate change is not the only, or even the most important, goal being pursued when these technological deployment projects are being developed. Mary MacDonald reminded me that job creation motivated the city of Toronto's initial interest in addressing climate change.[36] The mix of concern for the problem of climate change and desire to pursue economic development have been in the forefront of mayors' minds when considering climate policy since the inception of city action on climate change, and experiments that engage cities are well aware of the dual purpose that municipal actions on climate change are designed to serve. Dasha Rettew, manager of The Climate Group's Cities and Technology Program observes that cities are eager to participate in technology deployment projects precisely because "climate alone [is] not the only reason to do the project."[37]

Cities are thus motivated to engage with climate-friendly technologies for the cobenefits they promise. On the other side, corporations find cities and especially city networks ideal places to experiment with or pilot new technologies.[38] This is a crucial partnership because, as Dasha Rettew explains, most of the climate-friendly technologies just discussed are precommercial. While the technologies exist, the return on investment or value case is simply not there for corporations to roll them out commercially.[39] The technology is available but is not yet profitable in current market circumstances. Some experiments like The Climate Group and the Clinton Climate Initiative are working right at the rub of this chicken-and-egg problem. The market is not yet conducive to large-scale moves by corporations seeking relatively quick returns from selling technologies to cities, and the market cannot reach this state without a demonstration that the technologies have significant benefits for both cities and corporations. Thus implementing, financing, and reporting on pilot projects is a key objective for The Climate Group—strategizing in their networks to find means of gathering resources (financial and political) to experiment with precommercial technology in cities with an eye toward commercialization in the medium term.[40] The Climate Group's LED lighting project is a key example of this process. The corporate sector has developed this technology, but the demand for it was not initially there. Large-scale demonstration projects facilitated by The Climate Group have shown that the technology can be beneficial for cities, and demand is now growing.[41]

So the larger goal for some experiment initiators is working with cities and their networks to pilot projects and demonstrate proof of concept for wider application. Cisco is potentially in the midst of this transition at the moment. They began work on what would become the Connected Urban Development program as a corporate social responsibility initiative. They were looking at deploying information technology in Amsterdam, San Francisco, and Seoul to assist those cities in their sustainability planning, working to reduce their emissions from transportation and buildings. But the program did not remain a corporate social responsibility initiative. Nicola Villa, the current managing director for the Connected Urban Development program, noted that a presentation by Thomas Friedman at Cisco started the corporate leadership thinking about moving beyond corporate social responsibility. He recalls that Friedman's contention that there would be enormous sums of money to be made in addressing sustainability and climate change got their attention.[42] As the program developed:

> We sat down with the cities and our focus for the [first] 12 months was very much a CSR [corporate social responsibility] focus—how do we start developing technology architecture and services to tackle the climate change issue and reduce emissions across transportation, buildings etc. We started to then co-create and co-develop programs with the cities themselves. . . . And after about 10–12 months, we started to think about and see how those solutions and those architectures could become ready opportunities for Cisco along the way. So gradually over the last three years we moved our research from a CSR operation driven by the strategic consulting arm of Cisco into an incubator of technology solutions and services, which then Cisco would take to the business units and industrialize and spread in a commercial way. So the balance is tilted from only being CSR to being CSR *and* business oriented and therefore scoping opportunities which the company would start selling, not in the next 6 months, but in the next 2–3 years.[43]

Climate governance experiments are working at multiple levels of this interface of the corporate and municipal worlds, and a fascinating web of initiatives is forming. This is the very redundancy, or what Jan Franke, policy officer for EUROCITIES, described as "messy complementarity," expected in chapter 3 emerging in this activity cluster.[44]

At one level, implementing cities are often engaged with multiple experiments and participate in multiple municipal networks. For instance, Seattle was listed as a low carbon leader in September 2005 by The Climate Group for its efforts to improve energy efficiency and for "implementing transportation programs that reduce vehicle trips."[45] Yet part of Seattle's program on energy efficiency relies on corporations engaged with the Climate Wise experiment. So we have pursuit of technology deployment in cities recognized and facilitated by The Climate Group, but at least somewhat dependent on corporate activities shaped by a different experiment altogether.[46] This is not an isolated case. One-fifth of the C40 cities also engage either with The Climate Group (Chicago, formerly a member, Hong Kong, London, Mumbai, New York, Toronto) or the Connected Urban Development Program (Seoul, Madrid), and four (Chicago, Los Angeles, New York, and Philadelphia) of the five U.S. cities in C40 also signed the U.S. Conference of Mayors Climate Protection Agreement.[47]

Beyond overlapping cities in networks and experimental programs, the experiments themselves interact both cooperatively and competitively. There are a number of linkages across experimental initiatives and a range of different ideas about deploying technology in municipalities and their networks. Experiments are working in similar areas, and while their goals may be the same, the way they approach them can be very different, and the funds available for pursuing projects are limited. Competition is a natural outgrowth. In fact, Molly Webb and The Climate Group have some concern that cities are actually being bombarded by technology solutions at the moment and that funding and capacity need to catch up if the hoped-for market transformation is to come to pass.[48] Betsill and Bulkeley anticipated this dynamic when they argued that the challenge to successful municipal network action on climate change "stems from a lack of resources or powers to act."[49]

Deploying Technology in the Experimental System

The alliances, competition, cooperation, and parallel programs found in this activity cluster form a convoluted set of interactions, making it difficult to conceive of a linear narrative that captures how development of this activity cluster is structuring the experimental system. Tracing the origins and evolution of a particular relationship—the recently conceived alliance between the Cisco-initiated Connected Urban Development Program and The Climate Group—illustrates the complex connections in this activity cluster.

In late 2009, these two experiments announced their forthcoming alliance to much (internal) fanfare:

> Over the coming months, The Climate Group will reach out to its extensive global network of corporate partners, cities and states to develop the CUD Alliance and scope a new program to further advance its existing cities-focused work it currently delivers under the five-year HSBC Climate Partnership. Once the Alliance is in place, a new program—to be formally launched next year [2010]—will deploy urban demonstration projects in transformational technical areas such as smart connected buildings, smart transportation and smart grid.

> John Chambers, chairman and CEO, Cisco, says: "With urban areas contributing at least 60 percent of global carbon emissions, cities are ground zero for addressing climate change and environmental issues. The Connected Urban Development global community has successfully collaborated to test how innovative ICT solutions can manage environmental challenges. Cisco is committed to furthering this progress and asks other public and private sector partners to join us in this effort."

Steve Howard, CEO, The Climate Group, says: "Deploying smart low carbon technologies within world cities is central to unlocking energy efficiency at scale, and the transformation to a cleaner, greener, and more prosperous society and economy. The new CUD Alliance will align a global armoury of innovators, policy-makers, financiers and businesses necessary to pilot and scale carbon resilient systems, policies and practices for millions of citizens."[50]

The Connected Urban Development program is a voluntary actor experiment initiated by Cisco that counts cities as its implementing actors. It falls into the fourth category, technology development/deployment, focusing on reducing emissions from the use of information and communications technologies in a number of other areas within cities.

As noted, this particular experiment emerged from the corporate social responsibility division at Cisco in 2006. Like many corporations, Cisco was interested in maintaining a good reputation for environmental stewardship and looking for ways to contribute to the global response to climate change. The CEO of Cisco justified the program in this way:

> Our efforts to reduce carbon emissions are closely aligned with our corporate history of innovation, giving back, and taking risks, including socially responsible issues. Our belief is that we have a responsibility to help find ways to quickly reduce carbon producing activities more generally. What's more, we address the causes of these problems and work hard to implement systematic solutions using viable processes that will work for the long term and make our world a better place. I truly believe that this is our responsibility as a global citizen, and I'm excited about the role of networking in increasing the productivity of individuals, companies, and countries, while at the same time reducing the carbon footprint on a global basis.[51]

In the beginning, the Connected Urban Development program was considered to be a matter of good corporate citizenship.

Yet the focus of the program and the decision to deploy resources ($15 million for the first 5 years) in this manner was not an entirely internally driven interest. The program emerged from Cisco's relationship with another experiment initiator—the Clinton Foundation. In 2005–2006, the new Clinton group wanted Cisco to support their activities on climate change and beyond.[52] In early discussions, the Clinton Foundation began asking for more than monetary support—it wanted information and expertise about means of decoupling economic growth from climate damage.[53]

Both Cisco and the Clinton Foundation knew that everyone said that information technology has a role to play in addressing climate change, but the *how* part was still unknown. Villa recalls that Clinton challenged Cisco: "Can you, Cisco, founders of the internet," figure this out?[54] Both the Clinton Foundation and Cisco felt that cities were the place to begin, and Cisco began discussions with a broad number of cities. With the Clinton Foundation, they came up with some criteria for pursuing pilot projects. They wanted to initially work with heavily wired cities. A program in Amsterdam, Citynet, especially caught their attention. This program is bringing together telecommunications and real estate companies to enhance the connectivity of the city by linking 450,000 homes and businesses to a fiber optic network.[55]

This was seen as an ideal site to experiment with how cities could become more sustainable when they become more connected—how this connectivity could be harnessed to change the energy consumption and the transportation and all the flows that move through the city.[56] In addition to Amsterdam, Cisco identified San Francisco and Seoul as cities developing this next generation information technology infrastructure suitable for the Connected Urban Development experiment. These cities also met an additional selection criterion. The presence of local economic clusters (digital media especially) was crucial, the idea being that any application designed to work with the information technology network for enhancing sustainability and energy efficiency would first go through this cluster and then scale up to the city and then (hopefully) the rest of the world.[57]

The early development of the Connected Urban Development program (2007–2008) demonstrates how different experiments can provide different functions to create synergy in the experimental system. Cisco had the technological know-how to pull off the program, but the Clinton Climate Initiative brought what Villa calls "The Flag." The imprint of the Clinton Foundation "drove lots of energy . . . within Cisco and also outside of Cisco. . . . The cities, in many ways, were able to move ahead, much more quickly than they would have, because there was an urgency to report back to Clinton every 6 months and every year."[58] In addition, because the Clinton Climate Initiative had become the implementing organization for C40, there was access to a ready network of corporations and cities. Villa recalls that making connections through the Clinton Climate Initiative's network was an additional value added to the partnership beyond the brand.[59]

The Connected Urban Development program grew from this early beginning into a global pilot project that demonstrates the ways cities can reduce their emissions from their own ICT infrastructure and use ICT to reduce emissions and enhance economic growth in other areas. From its initial work in San Francisco, Amsterdam, and Seoul, it has expanded to numerous pilot projects in seven cities (Lisbon, Birmingham [UK], Hamburg, and Madrid joined the original 3 in

2008) that work on issues of transportation and energy system management through information and communication technology.[60]

But this is not just the story of a simple division of labor with two experiments partnering and using their respective strengths to push forward on an important project. It is more complicated than that, because at the same time that the Connected Urban Development program was emerging, the Clinton Climate Initiative and C40 were developing their own technology deployment programs for cities in the areas of building retrofitting, lighting and transportation systems.[61] The Clinton Climate Initiative programs do not duplicate the pilot projects pursued by the Connected Urban Development program, but there is overlap. These two experiments partnered in one area where their combined resources added value but pursue independent initiatives in others. This is a microcosm of the entire activity cluster. Multiple experiments are working in overlapping areas, forming linkages at times and working independently at others. What results is both a web of relationships and redundancy in the experimental system.

For instance, the EUROCITIES experiment recently introduced (December 2009) a "Green Digital Charter," which—much like the Connected Urban Development Program—asks cities to "implement a strategy to promote green connected cities, making the most effective use of ICT as a platform for the economic, social and environmental well-being of all citizens" through specific projects and partnerships.[62] Currently 22 cities have signed on to the charter (as of March 2010), including three Connected Urban Development cities—Amsterdam, Birmingham, and Lisbon.[63] Jan Franke, a policy officer with EUROCITIES, reports that the Digital Charter emerged from the European Commission's reaction to the SMART 2020 report put out by The Climate Group, noting that the report had a big effect in raising both awareness about and the visibility of the role of information technology in cities in addressing climate change.[64] This kind of loose connectivity and overlap is a key aspect of redundancy. It is not an exact duplication of efforts. Rather similar programs in the same geographic and political spaces are developed and run differently, though not entirely independently. For instance, EUROCITIES works directly with municipal political leaders and has targets. According to Jan Franke,

> The Eurocities charter is used as a political tool to continuously have the issue on the agenda . . . and raise appropriate support, whilst cities work with their private sector partners in local activities like the Connected Urban Development program and others to put technology on the ground.[65]

The SMART 2020 program approaches the issues from a different angle. It is a key piece of The Climate Group's work on cities and technology. The report that kicked it off, published in 2008 provides technical information to municipal and corporate leaders and it

> has quantified the direct emissions from ICT products and services based on expected growth in the sector. It also looked at where ICT could enable significant reductions of emissions in other sectors of the economy and has quantified these in terms of CO2e emission savings and cost savings.[66]

The original report arose from a partnership between The Climate Group and the consulting firm McKinsey, and it outlines how "ICTs could deliver approximately 7.8 GtCO2e of emissions savings in 2020."[67] Molly Webb recalls that The Climate Group had been engaging corporate leaders in a number of sectors (discussed in chapter 4) and a grant competition from the European Commission created an opportunity bring business innovation and environmental innovation together in the ICT sector.[68] The challenge The Climate Group has faced since the launch of the report is getting beyond the corporate social responsibility level and into the upper echelons of corporations where the initiatives outlined in SMART 2020 can be taken up and implemented broadly.[69]

Coming full circle, the new alliance between The Climate Group and Cisco's Connected Urban Development program should be seen in this light. Though Cisco had kicked off the Connected Urban Development program, they had not been able to involve other companies in their work or to make much headway into city networks.[70] The Climate Group's network of cities, subnational governments, and corporations was thus an ideal match. Cisco was involved in the research for the SMART 2020 report, and they were simultaneously considering what kind of partnerships, especially with NGOs, would allow them to scale up and expand their program.[71] The Climate Group wanted to transform SMART 2020 from a report into projects, and the Connected Urban Development program was looking for a platform from which to expand their projects.[72] The combined strengths provided a foundation for the alliance:

> and this is what we asked The Climate Group to do—to match-make between different companies and cities with similar needs and help put together projects . . . very large and globally relevant programs to be run on the ground.[73]

The Climate Group will bring its networking (corporate and governmental) to bear as a platform for advancing the aims of the Connected Urban Development program. It additionally sees a role for itself in the alliance in setting performance standards for the pilot projects and assessing their benefits.[74] By tracking them independently and bringing some consistency to the reporting, the alliance will be better able to make claims about the results of the projects.[75] The goal of the alliance between the Connected Urban Development program and The Climate Group is facilitating the transition from pilot programs, independently observed, into business opportunities that enhance the municipal and global response to climate change. In fact, in 2010 Cisco launched a "go to market" program called Smart and Connected Communities. "Cisco realized that by investing in a good CSR program, it discovered a huge commercial opportunity in a previously uncharted space."[76]

Problem Solved?

There is an enormous amount going on in this activity cluster, with multiple experiments working to deploy technology in cities. There is partnering and alliance formation, while at the same time competition and redundancy are evident. In telling the story of the triangle between the Clinton Climate Initiative, The Climate Group, and the Connected Urban Development program, this section paints a somewhat rosy picture of a budding division of labor among experiments and the ways different experiments can interact and cooperate to exploit synergies in their mutual quests for addressing climate change. Different experiments are bringing their different strengths and skill sets together in order to pursue common goals. Yet some relatively large challenges to expanding the influence of these projects are still looming, and even if an effective division of labor is emerging, it is not entirely clear how the challenges can be overcome.

For instance, chapter 4 discussed The Climate Group's LED lighting project, which provides a different approach than the LED program in the Clinton Climate Initiative. The Climate Group works with specific companies to facilitate large-scale testing of specific LED technologies in cities (e.g. Mumbai, Hong Kong, New York), while the Clinton Climate Initiative has a strategy of advising cities on the planning and management of outdoor lighting choices, providing logistical support for projects rather than advocating particular technologies.[77] This is not competition in the classic sense of a zero sum conflict over market share or resources so the emergence of alternative paths/approaches in not necessarily a problem. In fact, Dasha Rettew relays that in conversations with C40 the two initiatives have discussed the complementarity of their efforts.[78] Because the two groups have different strategies and different memberships, there was room for both to operate—they fill different, if competitive in the sense that they offer alternative modes of deploying technology, niches. There is an awful lot of work to do just in terms of deploying technology in cities, to say nothing of climate change writ large, and the optimal way forward, technologically or politically speaking, is simply not clear. In this sense, competition and redundancy, or multiple ways of implementing technological deployment programs, can be a good thing because it fosters innovation and provides multiple options for moving forward.

However, as mentioned, Molly Webb is concerned that cities are being inundated with potential technology proposals right now and that there is a lack of capacity to pick and/or implement technological systems.[79] This is especially troubling because funding for pilot projects, to say nothing of funding for full-scale implementation, is scarce on both the municipal and corporate sides of the equation. When I asked Nicola Villa how the prospects for scaling up looked, he admitted that it is a very difficult path.[80] Cities are strapped for cash. Few corporations are willing to "invest in incubation programs which would trigger revenue two to three years from now."[81] This echoes the very concerns The Climate Group is currently working to alleviate. Dasha Rettew notes that a good deal of their work is now involved with putting structures in place to facilitate discussions of financing.[82] There is a gap in what cities are ready to afford and what companies are ready to offer, and they are working on the process to close this gap.[83] The ultimate impact of this activity cluster may depend on whether they (or another initiative) are able to do so.[84]

Conclusion

The experimental activity aimed at deploying climate-friendly technology in cities is diverse and potentially promising, but significant questions remain as to what it demonstrates about the development of the experimental system and for the broader global response to climate change. A recent analysis of the possibilities and importance of household-level action in the response to climate change discussed the development of a "behavioral wedge" that can kick-start the climate response.[85] The dynamics discussed in this chapter are an apt analogue at a higher level of political organization. Not only can technology projects in city networks act as incubators of innovative ideas (technological and institutional), they have the potential to provide some of the momentum necessary to move toward decarbonization. Whether they do so will depend on whether this activity cluster develops as a division of labor, rather than a sorting mechanism, and whether it can generate a significant impact beyond city networks.

The potential tension between a division of labor trajectory and a sorting trajectory comes into focus in this activity cluster. The division of labor story is attractively illustrated in the new alliance between the Connected Urban Development program and The Climate Group and in the original partnership between the Connected Urban Development program and the Clinton Climate Initiative. Here we have explicit actions that exploit synergies made possible by functional differentiation in the experimental system taking place in a specific site. The Climate Group has a network of corporations and subnational governments, especially cities. The Connected Urban Development initiative has a set of voluntary projects developed for deployment in cities. Bringing the two together has the potential to leverage the strengths of both in order to scale up pilot projects—to move from 1 to 100 cities, in Nicola Villa's words.[86]

But will synergy win out over sorting? The development of so much activity in this realm actually increases the choices available to cities and corporations, unfortunately increasing the likelihood that sorting will triumph over synergy. In addition, the way these programs are being promoted, with the cobenefits of economic development and jobs inextricably intertwined with the goal of climate action, has the potential to switch the emphasis away from climate action. As long as the goals of making money or economic development dovetail with climate-friendly action, this cluster of activity and the multiple paths developing within it can (or at least have the potential to) do a great deal of good. But if the two diverge, it is not likely that climate change concerns will drive policy, and because a range of activities are being pursued in this activity cluster, choosing ones that fit with economic interests rather than ones that are effective in combating climate change becomes a real danger.[87]

Because this activity cluster has so recently emerged, it is natural that a number of activities are being tried out to see which stick—economically, politically, and in terms of climate action. This overt experimentation has the potential to build friction in multiple dimensions. Corporate friction can emerge as firms compete to participate in these new technological markets. Friction between cities and corporations may arise as multiple building standards, energy efficiency, energy sourcing, and transportation policies emerge across municipalites. As more and more cities engage in this activity cluster, the number of different municipal policies will multiply, increasing the incentives and constituencies behind broader and uniform policies. Friction between cities and other governmental levels could also materialize as cities push for greater control over climate related policy areas (transportation, urban planning, and energy production) and subnational and national governments need to deal with markets that emerge for renewable energy and other technologies.

Further, in building the infrastructure for cities to respond to climate change, these experiments may be providing the means for a broader responses—the technologies, institutions, policies, and knowledge necessary to scale up to regional, national, and global responses. The overriding goal of the initiatives in this activity cluster is market transformation. If the programs can scale up effectively and find the funding to implement them broadly, they may be the catalyst that pushes the economy toward a more climate-friendly footing.

Assessing whether this is a likely outcome is another question. The experimental system and its clusters of activities are emerging; they are not fully functioning.

Right now they are too small and too new to have much of a tangible impact on global climate change (or even local economic development). They are thus difficult to study in the sense of providing definitive perspectives on their functioning, politics, and relevance. Right now they are hamstrung by lack of funds, and the path to scaling the various projects is as yet unclear. But there are signs of hope and momentum. Nicola Villa claims that

> the businesses and investors and the NGOs are really coming together on operational programs. There is quite a bit of energy and quite a bit of movement in the business world, independently of what happened in Copenhagen which needs to be taken into account.[88]

There is much to do financially and in terms of the regulatory world to make the technological development/deployment projects more than pilot programs. The key question is whether these initiatives can provide enough friction and smoothing to lead or at least contribute to the global response to climate change.

The analysis in this chapter uncovers how relationships and networks are self-organizing in this specific activity area or governance site—municipal technological deployment—in a way that is facilitated by the functional differentiation evident among experiments. These clusters are developing along a trajectory that may see them be a source of enormous friction in political and economic systems that can in turn serve as a source of disruptive change. The development of denser networks and more alternative paths of implementation will enhance this friction generating process, but it is unclear if the resources and capacity are yet in place to scale up these programs.

6

Constructing Carbon Markets

The UN invited us to present a paper at the Earth Summit in
Rio de Janeiro in 1992 on the potential for a global emissions
trading system. The idea was received with skepticism,
but it became clear to me that there was a need to build
institutions for this new commodity—carbon—which
had the potential to be the biggest one in the world.
—Richard Sandor, cofounder of the Chicago
Climate Exchange (summer 2007)

Carbon markets—the notion of putting a price on greenhouse gas emissions by commodifying, buying, selling, and trading them—have gained significant momentum in the last few years.[1] Given the prevalence of "market mechanisms" shaping almost all facets of life, and especially in the environmental area, the inclusion of market mechanisms in the global response to climate change appears entirely natural. Yet in some very important ways, addressing climate change through the creation of carbon markets is a grand experiment in and of itself. Experimentation implies trying something to see if it will work (with hopefully an educated sense of what might work) and in this sense, initiating a market for greenhouse gas emissions is a huge experimental endeavor. There are clearly big questions about the effectiveness and even morality of carbon markets as well as debates over how markets should be designed and function.[2] The carbon market experiment involves the construction of a brand new commodity and market from scratch and a cluster of (strictly defined) climate governance experiments has emerged to participate in this building and the exchange of commodities that results.

This chapter thus examines experimentation on two levels. The first is the carbon markets themselves, and I begin with a brief overview of their logic, history, and current manifestation. What was conceived in the Kyoto Protocol as an integrated global carbon market with both allowance (cap and trade) and credit (offset) aspects has fragmented into multiple and only loosely connected cap and trade systems and credit markets. The second is the role climate governance experiments play in carbon markets. A cluster of experimental activity has emerged around the development of carbon markets. I focus on how infrastructure building experiments have emerged to set standards in the voluntary credit markets and how accountable actor experiments have engaged in the design and functioning of cap and trade systems. In contrast to the deployment of technologies in city networks, experimental activity in the carbon markets is inextricably linked with traditional governance mechanisms. Symbiosis of experimental and traditional governance is clearly apparent. Because of this interdependence of governance systems, the friction generating and smoothing functions provided by the experimental world may be even more significant in this area.

The Carbon Market Experiment

The philosophy behind and promise of carbon markets is that by making greenhouse gases into commodities—putting a price on them—we can harness the power of markets to achieve environmental goals, hopefully in the cheapest possible manner. From this perspective, climate change results from a massive market failure caused by negative externalities. Electricity producers, corporations, even individuals are not paying the full costs of using fossil fuels. Instead, these costs (the environmental damage from climate change) are externalized onto the whole world. Putting a price on greenhouse gas emissions is a way to internalize these costs and drive investment away from activities that produce greenhouse gases. The logic behind carbon markets is to commodify emissions in order to price and reduce them (see box 1). The last 15 years have been a fascinating case study in the development of just such a market (or markets, really).

The notion of carbon markets emerged as a serious policy option in the negotiations over the Kyoto Protocol, with a good deal of the debate at Kyoto surrounding the place market mechanisms should have in the global response to climate change.[3] They were part and parcel of the flexible mechanisms the United States advocated for, arguing that market mechanisms would control the costs of reducing greenhouse gas emissions. But as mentioned in chapters 3 and 4, the idea of using market mechanisms in service of environmental goals is a familiar motif in U.S. environmental policy and in the OECD writ large.[4] Early knowledge networks and experience with emissions trading to address acid rain and air pollution in the United States laid the foundation for adapting this policy tool

---

Box 1 **Carbon as Commodity**

In some ways, carbon markets are like any other commodity markets. Suppliers have a commodity that those on the demand side want to consume, and the market determines the price by matching levels of supply and demand. But carbon as commodity differs from other commodities in some significant ways. In both credit and allowance markets, the commodity is a set amount of greenhouse gas emissions (most often a metric ton). They can be created in two ways, and both require some sort of policy decision to make a commodity where there was nothing but air before. In the allowance market, the commodity itself is created by the policy of instituting a cap on emissions for a group and dividing the cap into permits to emit greenhouse gases. When the CCX puts a ceiling on emissions for its members (collectively 1% below a calculated baseline per year) or when the RGGI caps the amount of emissions that electrical power generators are allowed, they are literally creating a commodity—permits to emit a certain amount of greenhouse gases. In the credit market, the commodity is a promise to not emit greenhouse gases. Offset producers reduce their emissions and sell the difference between a baseline and actual emissions—emission credits. In both cases, then, the commodity becomes a license to emit a certain amount of greenhouse gases—i.e., the buyer of offsets or holder of permits can emit that specified amount of greenhouse gases. But in contrast to traditional commodities—minerals, grains, livestock, and so on—nothing tangible is being produced or changing hands. Instead, what is changing hands are promises and permissions. A significant challenge in both kinds of markets is devising procedures for measuring, verifying, and reporting emissions and emissions reductions so that the "commodity" has some integrity.

---

to the larger problem of climate change.[5] In addition, creating carbon markets suited the liberal environmental ethos that permeates the global response to climate change.[6]

The original vision was to have an integrated global carbon market associated with the Kyoto Protocol consisting of a global cap and trade system and a global offset system that engaged the two types of actors participating in the multilateral negotiations—Annex I countries that have emission reduction commitments (developed countries) and non–Annex I countries that lack such commitments (developing countries).[7] The cap and trade system was to engage Annex I countries and facilitate their achievement of the negotiated emission reductions. Along with a cap and trade system, the Kyoto Protocol laid out a complementary credit or offset market. In credit markets, actors undertake activities or projects to reduce greenhouse gas emissions from some baseline (plant trees, change land use, invest in energy efficiency or renewable energy, etc.). The reductions are turned into emission credits—tons of greenhouse gases reduced and not emitted—that can be sold to consumers who seek to manage their greenhouse gas emissions (either voluntarily or by mandate). The Kyoto Protocol initiated two credit markets that could be used by Annex I countries to meet their emission reduction commitments. The Joint Implementation initiative was for offsets produced in Annex I countries (especially transitional economies in central and eastern Europe). The CDM was negotiated as a way for developing countries to participate in the carbon market—producing credits that could be sold to entities with reduction commitments, simultaneously advancing sustainable development goals.[8] Whatever the original intent, the envisioned integrated global carbon market has given way to fragmented global carbon markets.

MULTIPLE CAP AND TRADE SYSTEMS

Rather than a single cap and trade system under the Kyoto Protocol, "stagnation in the multilateral negotiations and the withdrawal of the United States in 2001 led to significant fragmentation in this 'global' market."[9] Currently it is possible to identify 32 distinct venues where cap and trade has been proposed, is under development, or has become operational (see table 6.1).[10] These venues emerged in diverse places (primarily in the industrialized world) and political jurisdictions (from the global to the local) and include both the public and private sectors.[11] What was devised as a centralized cap and trade system to serve the ends of the multilateral governance system has become a fragmented carbon market with multiple emissions trading systems, in which many questions about how to undertake emissions trading are contested.

With multiple cap and trade venues come multiple visions and designs for pursuing emissions trading. At the most basic level, there is no international consensus about which actors should undertake cap and trade policies. Venues can be found across the spectrum: international EU Emissions Trading System), national (Japan, New Zealand), subnational (RGGI, WCI), and private (CCX). Of particular note is that half of the cap and trade venues identified are being or were designed by subnational authorities (alone or in concert). All but one of these is located in either North America or Australia, where subnational discussions of general and cap and trade are justified as a necessary response to federal inaction.

Diversity is found not only in the actors taking up cap and trade as part of the response to climate change but also in the design of cap and trade systems themselves. The general idea of cap and trade has proliferated, but its operationalization

Table 6.1 **Emissions Trading Venues***

Venue	Initial Discussions	Trading Begins
British Petroleum	1997	1998
New Jersey	1998	
Shell	1998	2002
United Kingdom	1998	2001
Canada	1998	
Norway	1998	2005
New South Wales	1998	2003
Denmark	1999	2000
European Union	1999	2005
Chicago Climate Exchange	2000	2003
Switzerland	2000	2008
Massachusetts	2001	
New England Governors	2001	
New Hampshire	2001	
NAFTA	2001	
Japan	2002	2005
US Congress	2003	
RGGI	2003	2009
Australian States and Territories	2004	
Western Climate Initiative	2004	Set to Begin 2012
Oregon	2004	
New Mexico	2005	
California	2005	
Illinois	2006	
Australia (Federal)	2007	
Florida	2007	
Midwestern Greenhouse Gas Reduction Accord	2007	
New Zealand	2007	2008
South Korea	2007	
Ontario-Quebec	2008	
Tokyo	2008	Set to Begin 2010
PEMEX (Mexican Petroleum)	2009	

Note: Bold indicates that a venue is currently operational (i.e., actively trading).

*Source: Betsill and Hoffmann 2011.

is far from uniform. There is considerable variation in proposals for how a cap and trade system should be designed across the policy venues. Important design elements and debates include:[12]

- *Which greenhouse gases to cover:* Some venues limit coverage to CO_2, which is easier to monitor, and others cover CO_2 plus other gases (often all of the six greenhouse gases listed in the Kyoto Protocol), which gives regulated entities greater flexibility to achieve emissions reductions.
- *Which economic sectors to include:* The majority of the systems regulate multiple economic sectors, but some target electrical generation.
- *Mandatory versus voluntary systems:* This was initially a large source of debate, though no venue initiated after 2002 has proposed a voluntary arrangement.
- *Allocation of permits:* One of the most contentious issues in designing cap and trade systems is how permits are distributed to regulated entities—through free allocation or auctioning. The temporal trend is toward auctioning. The preference for auctioning since 2003 likely reflects lessons learned from the EU experience, in which free allocation along with an overallocation of permits led to a dramatic collapse in permit prices. Since then, policy-makers tend to prefer that at least some permits be auctioned, in order to send a clear price signal and to avoid charges of windfall profits.
- *Use of offsets:* The vast majority of venues that get to the design or operational phase allow regulated entities to use credits purchased from offset projects to meet their commitments. Where offsets are allowed, there is wide variation in the specific rules governing their use. Some venues, such as the EU Emissions Trading System, allow the use of CDM credits (although often on a limited basis in terms of sectors or percentage of total emissions covered). Others place specific limitations on the geographic locations where offset credits may be generated. For example, in the RGGI, the majority of offset credits must come from projects within the United States unless a specified price trigger is met.[13]

Cap and trade has become a frontline, if controversial, policy in the global response to climate change. Drawing on earlier experience with emissions trading, it was nonetheless innovative to include emissions trading as a mechanism in the Kyoto Protocol. Similarly, it was experimental for a number of nation-states to pick up the mechanism and discuss or implement it nationally, either as a move toward Kyoto compliance (UK, Denmark, and EU) or as a stand-alone mechanism (Australia). Yet an important turning point for the development of cap and trade systems occurred in 2000–2001. In early venues, proposals for cap and trade were justified in terms of gaining practical experience with trading in anticipation of an international trading system centrally organized through the Kyoto Protocol. This changed around 2001, as venues in the United

States and Australia began to consider setting up alternatives to the Kyoto system and as uncertainty grew about the long-term viability of the Kyoto Protocol following the U.S. withdrawal. In this way, the development of the cap and trade mechanism parallels the development of the experimental governance system itself—flourishing in the aftermath of stalemate in the traditional multilateral governance system. As we will see, cap and trade has also emerged as a key mechanism attracting climate governance experiments.

OFFSET MARKETS

The global credit markets have also significantly expanded since they were first conceived in the Kyoto Protocol negotiations. The CDM is functioning and is actually a (relatively) thriving market producing greenhouse gas credits that can be used in cap and trade systems (mainly in the EU trading system that emerged in 2005).[14] It is known as the compliance offset market because CDM credits can be used to fulfill mandated emissions reduction commitments (for instance in the EU emissions trading system). Yet a voluntary offset market has emerged alongside the CDM, catering to nonregulated entities. This has been especially important in North America, because the U.S. withdrawal from the Kyoto Protocol removed a huge potential source of demand for CDM credits.

Developing credits—specified quantities of avoided emissions that consumers would like to obtain in order to "offset" their own emissions of greenhouse gases—that can serve as a reliable commodity is not a simple matter. Those providing the commodity are essentially giving a promise to not emit that specified amount of greenhouse gases so that the buyer can consume the commodity by emitting that specified amount of greenhouse gases. Making this work requires devoting significant energy to measuring, verifying, and tracking credits to ensure that the exchange of credits has integrity that is both economic—avoiding fraud and double-counting of credits—and environmental—ensuring that emissions are actually reduced.[15] There must be systems in place to assure the integrity of credits. This means accounting for the additionality (credit is only granted for emissions reductions that would not have occurred in business-as-usual circumstances) and permanence of the emissions reductions. In addition, it is necessary to establish clear ownership of credits and to ensure that they are only "consumed" once.

In the CDM, these tasks fall to the executive board of the CDM, which was created and is backstopped by the multilateral treaty-making process and the secretariat of the UNFCCC. It is more complicated in the voluntary sector because there is no single legal authority to make and oversee the rules. The voluntary carbon market is actually a complex web of relationships. Figure 6.1 provides a simplified schematic of the infrastructure that is developing for the

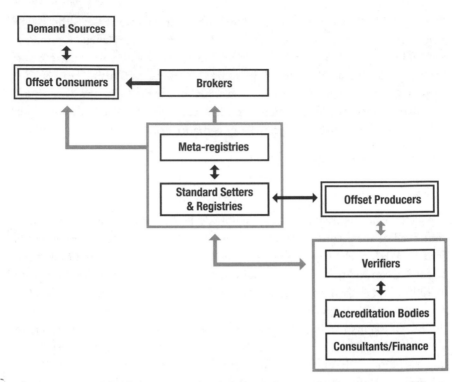

Figure 6.1   Infrastructure for the Credit Market

production and exchange of offsets. This is a general schematic that captures the essence of this process whether we consider the voluntary or regulated sectors (though I will restrict this discussion to the voluntary sector). Figure 6.1 does not include every relevant actor, but it does provide a sense of the players connected most directly with the production and consumption offsets and the interactions that create and move the market—a brief digression to explain the workings of the credit markets is warranted here.

Beginning on the consumption side (upper left), there are generally three types of consumers that differ on the source of demand for offsets. First, regulated entities are in the market for offsets. The source of regulation is obviously diverse. Some have commitments they must meet because of national regulations or because the nation-state in which they operate has Kyoto Protocol commitments. Others are participating in established cap and trade systems. Examples would include the power generation sector in the RGGI or corporations that have voluntarily taken on explicit and binding reduction emission reductions commitments in the CCX. Mary Grady, directory of registry and membership services at the American Carbon Registry, further notes that corporations that *anticipate* being regulated in the future are also engaging in the market.[16] For instance, in May 2009, Duke Energy became a lead investor in a forest offset project, now approved by the American Carbon Registry. A significant motivation for this investment in offset production was anticipation of regulation:

> Duke Energy expects the federal government will soon implement an economy wide cap-and-trade program to control carbon dioxide emissions. The company also believes that high-quality, verifiable carbon offsets derived from reforestation efforts will play an important role in such a program.[17]

Second, a number of corporations or other organizations (cities and universities for instance) and even individuals consume offsets for environmental prestige or corporate social responsibility reasons. Given the lack of a global treaty or significant U.S. federal response, combined with significant interest in climate change from consumers, stakeholders, and shareholders, some organizations have decided to move toward carbon neutrality in the absence of regulations demanding it in order to showcase their environmental commitments and credentials. Mary Grady remarked that Google was a prime example of a corporation purchasing offset credits through the American Carbon Registry and "retiring" those credits in order to pursue carbon neutrality even without the specter of regulations prompting them to do so.[18]

Finally, speculators are playing in carbon markets. The carbon market is already growing (while on the voluntary credit market *only* $400 million worth of transactions took place in 2008, the secondary market for CDM credits—trades beyond the initial purchase of credits—climbed to $26 billion in 2008),[19] and some financial institutions are betting (at least a bit) that offset markets will be a major sector in the years to come. Grady noted that financial players were a significant category of participants in the American Carbon Registry.[20] The 266 registered account holders on the Climate Action Reserve include 52 entities described as trader/broker/retailer. and these include such financial industry heavyweights Barclays Bank, Citigroup, JP Morgan, Morgan Stanley, and Royal Canadian Bank. FurtherPoint Carbon's[21] "Carbon Market Insights America" conference in November 2009 clearly demonstrated that large financial institutions are interested in the carbon markets. Informational booths were maintained by JP Morgan and Morgan Stanley, and conference sessions were as often geared toward financial sector concerns or opportunities as toward emission reduction policies and projects.[22]

Demand for offsets presupposes a prior step in the development of this market—entities need to know what their emissions are before they can even decide whether or how to participate in the credit markets. Before one can take steps to reduce emissions, one must account for those emissions. A number of methodologies have emerged in the last decade to do just that. Often these methodologies are associated with carbon registries like the Climate Registry introduced in chapter 4—an accounting program for organizations to keep track of their emissions—or with a trading system itself—like the CCX. Some, like the Carbon Registry, have strict guidelines on the measurement and verification of emissions. Others are more guidelines for self-reporting. Organizations engaged in the carbon markets measure their inventories and have them verified by third parties so that the emissions can be registered and a baseline and reduction and offset activities can be tracked. A significant motivation for the development of these registries has been precompliance action—getting organizations ready for regulations and allowance markets that are being or may be developed.

Demand for offsets is being met by a variety of producers and projects, and the number and diversity of projects that can now be used to commodify emissions reductions is staggering. They generally fall into one of these four general categories:

- *Land use and forestry activities:* Reforestation or avoided deforestation (tree planting or avoided tree cutting) are crucial in this area. In addition, the CCX has also encouraged the development of offset credits from new farming techniques, such as no-till farming.
- *Alternative energy and energy efficiency projects:* Credits are developed by investing in wind, solar, and other types of alternative energy that can demonstrate direct replacement of fossil fuel based energy production.[23]
- *Landfill and waste management and methane conversion:* Using methane from various sources (mines, landfills, biomass, waste) to generate power and electricity.
- *Managing ozone depleting chemicals:* This has been a main source of offset credits generated through the CDM. It involves destroying ozone depleting chemicals, which are gases with very high global warming potentials that are also regulated in the ozone depletion regime (these were the original replacement chemicals for the CFCs that were phased out under the Montreal Protocol).

This is where things get interesting and complicated—measuring and verifying the credits. It is relatively simple to begin an offset project. The hard part is turning the project into a product—commercializing efforts at reducing emissions. The explosion of offset providers in the early 2000s generated a great deal of concern about the integrity of the markets, because offsets were being developed and sold with little in the way of oversight over what was being purchased.[24] This was a big concern in the retail market but caused even more significant worries among those committed to the development of carbon markets as a key component of the global response to climate change. Without the ability to identify what counted as a ton of greenhouse gasses and what counted as a ton of avoided emissions, the whole carbon market (voluntary or regulatory) would collapse (taking any hope of achieving emissions reductions through this mechanism with it). Taking a cue from the Kyoto process, which developed standards within the CDM program, standard setters began to emerge in the voluntary offsetting markets as well, in order to "bring order to the wilderness."[25]

The key to the credit market is producing promises to not emit and reduce greenhouse gases that are verifiable, permanent (not emissions that will resume in subsequent years), additional (not something one would do anyway or something one is required to do), and clearly owned. In other words, the credit market depends on knowing that an identifiable ton of greenhouse gas is not going into the atmosphere that otherwise would have, that the seller of that ton has a right to sell it, and that it only gets "consumed" (retired or turned in to meet requirements) once. Standard setters (e.g., American Carbon Registry, Climate Action Reserve, Voluntary Carbon Standard, and Gold Standard) have developed or approved a number of specific methodologies for projects to ensure that they meet these more general criteria. But standard setters do not go to each project to see if their standards are being met. Different kinds of actors—validators and verifiers—have emerged to perform this task. These actors take in the information produced by offset project developers and check to see if the offset project proponents have met the offset standard setters' requirements. They report back to the standard setters, who then issue serialized offsets—uniquely numbered certificates certifying a specified number of tons of reductions.

Once created and verified, offsets are listed on offset registries. Some standard setters, like the American Carbon Registry, maintain their own registry and accounts for the transactions of the offset credits they approve. Others, like the Voluntary Carbon Standard and the Climate Action Reserve, use metaregistries—companies that do carbon accounting—to track the offsets that they approve. Metaregistries often provide the software platforms for registries (one company, APX, provides the software platform for the Climate Action Reserve, for instance) and can accommodate transactions that take place under a number of different offset standards. They receive information on the serialized emissions reductions issued by offset standard setters and keep track of them for offset purchasers. If we think of offset standard setters as producing different kinds of carbon currency—that is, a Voluntary Carbon Standard ton is distinct from a Climate Action Reserve ton (even though functionally a ton is supposed to be a ton), then we can think of metaregistries as allowing organizations to keep track of their accounts in the different currencies.

But of course someone has to watch the watchers. So accreditation bodies have begun to work in the offsetting sector to train and accredit verifiers. In addition, someone has to pay for this activity. Project developers looking for assistance in meeting standards or getting financing for their projects turn to consultants and financial players. A new industry is emerging to facilitate the matching of projects and standards in a fruitful manner.[26] This can be complex, because some consultants not only assist developers who know they want to participate in the market but also work to entice project developers—advertising the possibility of monetizing privately held forests, for instance.[27] In addition, the finance players can be on both sides of the equation (offset consumers and investing in offset producers projects). A number of companies now provide finance for offset project developers and are getting their returns in offsets rather than money.

Offset consumers purchase and track their offsets directly from producers, through the offset and metaregistries, or through brokers/retailers. They then consume the offsets by turning in the permits to regulators, trading them on an emissions trading system, or reporting them in corporate neutrality reports, satisfying the source of demand. Of course, the market is not restricted to direct transfers of credits. The more lucrative aspect of the credit market (and more problematic, according to some critics) comes from secondary trading of credits. In fact, futures and derivative markets have already emerged for carbon offset credits.[28]

Akin to the global experience with cap and trade mechanisms, carbon credit markets have expanded beyond the original vision embedded in the Kyoto Protocol. Carbon credits are being produced around the world, to varying standards, as part of the process of commodifying greenhouse gas emissions and providing opportunities for those seeking to manage their own emissions to do so cheaply and without necessarily fundamentally altering their own activities. The idea of carbon credit markets is experimental in and of itself, and making them function has been a key area of activity for climate governance experiments.

The Experimental Carbon Market Cluster

As carbon markets have developed over time, they have become both a source of experimentation (i.e., experiments have emerged to participate in carbon markets) and a cluster of experimental activity. They serve as a major attractor for climate governance experiments, and initiatives from all four governance models are playing roles in various carbon markets. Of the 58 experiments, 16 are directly involved with credit and allowance markets in some capacity.[29] Networkers are spreading best practices about the design and functioning of markets. Infrastructure builders are working on the measurement, verification, and accounting of emissions and emission reductions in credit markets. Voluntary and accountable actor experiments are engaged in designing trading systems for cap and trade mechanisms beyond the Kyoto Protocol. Experiments are undertaking a wide variety of activities to make carbon markets a working reality. Table 6.2 gives

Table 6.2 **Experiments in the Carbon Market Cluster**

Experiment	Type	Activities
2 Degrees	Networker	Carbon Compliance and Management working groups
American Carbon Registry	Infrastructure Builder	Offset registry and standard setting
California Climate Action Reserve (Now Climate Action Reserve)	Infrastructure Builder	Offset and footprint registry, Standard setting
Carbon finance Capacity Building Program	Voluntary Actor	Advises municipalities on how to monetize emission reduction programs.
Carbon Fix	Infrastructure Builder	Offset standard setter
Chicago Climate Exchange	Accountable Actor	Cap and trade system, also sets standards for offset projects.
Climate, Community, and Biodiversity Alliance	Infrastructure Builder	Offset standard setter—social and biodiversity standards for land-based carbon offset projects
Global GHG Register (defunct)	Infrastructure Builder	Carbon footprint registry
International Climate Action Partnership	Networker	Carbon market design and advice
Midwestern Greenhouse Gas Reduction Accord	Accountable Actor	Regional cap and trade system

continued

Table 6.2 (**continued**)

Experiment	Type	Activities
Ontario-Quebec Provincial Cap-and-Trade Initiative (defunct—provinces joined Western Climate Initiative)	Accountable Actor	Regional cap and trade system
Regional Greenhouse Gas Initiative	Accountable Actor	Regional cap and trade system
The Climate Group	Networker	Initiated the Voluntary Carbon Standard (offset standard) with the World Business Council on Sustainable Development.
The Climate Registry	Infrastructure Builder	Carbon footprint registry
Western Climate Initiative	Accountable Actor	Regional Cap and Trade System
World Business Council on Sustainable Development (and Voluntary Carbon Standard)	Networker	Initiated the Voluntary Carbon Standard (offset standard) with The Climate Group

a broad look at the experiments involved in this cluster of activities. But beyond being an attractor for climate governance experiments, carbon markets also represent an intersection site where the traditional governance sphere dominated by nation-states/UN process and the experimental governance system come together. In fact, it is safe to say that the global carbon market, such as it is, would look entirely different (and might not even exist) without climate governance experiments.

The explanation developed in chapter 3 for the emergence and shape of the experimental system is again borne out by the advance of experimental activities in carbon markets. The experiments in this activity cluster found their motivation broadly in the sequenced signing and failure of the Kyoto Protocol. Initially, the Kyoto Protocol negotiations were a catalyst that sparked innovation around carbon markets. The prospects of a global cap and trade system with a linked global offset system spurred a number of actors to prepare for and, in the case of the CDM, start participating in the global carbon market.

Many early emissions trading systems were justified in terms of gaining practical experience with trading in anticipation of an international trading system centrally organized through the Kyoto Protocol.[30] New Jersey and the Netherlands signed a Memorandum of Understanding in 1998 agreeing to cooperate to design a credit banking system for adoption in the multilateral regime.[31] British Petroleum and Shell implemented internal corporate trading systems so that they would be well positioned to participate in an internationally organized trading system.[32] Jon Skjaerseth and Jorgen Wettestad argue that "the [European] Commission expected international trading to become operational under the climate regime from 2008" and that trading within the European Union would be a way to gain practical experience in preparation.[33]

Similar dynamics were evident in the offset markets as well. The idea of the CDM—that one could produce a commodity for sale by reducing carbon emissions—spurred a number of activities aimed at laying the groundwork for markets beyond the Kyoto Protocol. The American Carbon Registry finds its roots in this drive—to build the infrastructure for carbon markets, which requires measuring, verifying, and reporting greenhouse gas emissions and reductions. In addition, because even the envisioned globally integrated carbon market would not regulate every entity, there was a push to develop voluntary credit markets to meet the demand coming from unregulated actors who nevertheless wanted to offset their greenhouse gas emissions.[34]

The failures of the Kyoto Protocol process provided additional, and perhaps more significant, impetus for experimental action. Subnational governments in North America explicitly turned to allowance markets as a way to address climate change in the absence of federal leadership and to spur national and international action. Franz Litz, one of the principal designers of RGGI, notes that the regional cap and trade systems were designed to send a political message about the practicality of addressing climate change to recalcitrant national governments in North America. They emerged as a reaction to the stalemate in traditional climate governance and were explicitly intended to make stakeholders want a federal program by setting up a patchwork carbon market in North America.[35] Statements from the governors that convened the WCI echo this sentiment:

- Janet Napolitano (Arizona): "In the absence of meaningful federal action, it is up to the States to take action to address climate change and reduce greenhouse gas emissions in this country. Western States are being particularly hard-hit by the effects of climate change."
- Arnold Schwarzenegger (California): "This MOU sets the stage for a regional cap and trade program, which will provide a powerful framework for developing a national cap and trade program. This agreement shows the power of states to lead our nation addressing climate change."
- Bill Richardson (New Mexico): "With this agreement, states are once again taking the lead on combating global climate change—while Washington, D.C. sits on its hands. This historic agreement signals our commitment to tackling the problem head-on at the regional level and building on efforts in our individual states."
- Ted Kulongoski (Oregon): "Today's announcement shows how the West continues to lead the way in addressing the most pressing environmental challenge of our time. Together, we are putting ourselves on a path to reduce greenhouse emissions and create a sustainable energy future—a model and example for rest of the nation."
- Chris Gregoire (Washington): "We have all seen the science and we must increase our efforts to respond. We must implement what we all have put in place, and work together to develop a regional market approach. Together, we can reduce our climate pollution, grow jobs and move toward energy independence."[36]

In addition, while the CDM initiated the idea and practice of a credit market, it was plagued by bureaucratic issues involved with approving projects, hamstrung by the lack of U.S. participation, and challenged with legitimacy issues.[37] Toby Janson-Smith, former director of the Climate, Community, and Biodiversity Alliance and a participant in the development of the Voluntary Carbon Standard, recalled that the UN political process resulted in the CDM framework not being very workable for forestry projects.[38] David Antonioli, the current CEO of the Voluntary Carbon Standard, additionally noted that there was a growing demand for credits coming from entities not regulated (i.e., corporations in the United States not subject to the Kyoto Protocol) who wanted to participate in the credit markets for corporate social responsibility reasons or to catalogue their early actions for credit in the eventuality of regulations.[39] A voluntary carbon credit market emerged into this vacuum. However, it was a bit of the Wild West, with offset producers popping up literally all over the place selling dubious offsets. Janson-Smith observed that a lack of transparency and integrity in carbon credits was significantly holding back the market, and Antonioli observed that demand for credits was being met with offsets that lacked reliability and credibility—consequently, worries grew about a backlash against offsets and the carbon markets writ large.[40] A number of experimental actors—registries and offset standard setters—were initiated to address this problem explicitly and began to govern the credit markets in important ways.

Yet experimental activity in the global carbon market has not just been about filling holes. As it has developed, it is becoming clear that the experimental system has generated a momentum of its own, thriving and even setting the agenda for the development of carbon markets through more traditional means of governance—national policy and multilateral treaty-making.[41] The Memorandum of Understanding that officially launched RGGI in 2005 makes this explicit, noting in its preamble that the signatories "wish to establish themselves and their industries as world leaders."[42] And more recently initiated cap and trade systems are often justified as a means for achieving local (rather than global) policy objectives like economic and technological development and local emission targets. Instead of looking to *join* a global cap and trade system, policymakers often talk about *creating* an international trading system from the bottom up by linking markets organized in different political jurisdictions. The three major regional cap and trade venues in the United States have recently begun meeting to discuss linking their efforts and presenting a united front in talks about the potential U.S. federal legislation.[43] Rick Saines of Baker and Mackenzie argued in front of a Point Carbon conference audience that linkage of domestic emissions trading systems was likely to continue *regardless* of the status of international treaty-making.[44]

Further, the energy and activity in the voluntary carbon credit markets is being incorporated into traditional governance mechanisms. Standards set for voluntary offset projects and methodologies important to designing voluntary carbon footprint registries are actively sought by national governments when they consider how to design and regulate carbon market activities. Indeed, "as markets developed over the last decade, an elaborate set of governance structures has emerged to address concerns around measuring and accounting for emissions and offsets, and tracking the permits and credits across the carbon market."[45] Much of this activity, in addition to actual exchange of greenhouse gas permits and credits, is happening in and because of the experimental system. Examining the experimental role that cap and trade systems and the voluntary credit markets are playing illuminates just how intertwined the experimental and traditional governance systems are and the way this relationship is shaping both the development of the experimental system and the broader global response to climate change.

EXPERIMENTING WITH EMISSIONS TRADING

It is telling that even while momentum behind action on climate change ebbs and flows, *none* of the venues listed in table 6.1 ultimately abandoned the policy entirely (NAFTA and Canada have come the closest, but both will likely look for a role in any potential U.S. federal cap and trade system, which itself appears may be a challenge to develop given legislative and political setbacks in 2010). It does seem clear that the international community has moved away from the assumption that cap and trade should be governed through the multilateral treaty process. Instead of a single global cap and trade mechanism situated in the multilateral treaty regime, it is more likely that any global cap and trade market would have to emerge from linking systems so that permits can be traded. This perspective was widely acknowledged in trading discussions at the UNFCCC COP at Copenhagen conference.[46] Economists, financial experts, and international lawyers are now focused on the technical aspects of creating such linkages, suggesting harmonization when possible or explicit contracts about the conditions under which permits generated in one system might be recognized in another system.[47] Linking will be no mean task. Beyond the rough political terrain in general for carbon markets in 2011, the design of individual systems reflects political compromises, so it is reasonable to expect that any efforts to link systems may revive political debates as stakeholders seek to ensure that their interests are promoted by linking. Proposals for linking trading systems to create a global cap and trade market must anticipate and account for the political debates likely to arise from these dynamics and consider options for resolving such debates.[48]

Climate governance experiments will have a significant voice in these debates, and the relevant focus here is on the role(s) the experimental governance system is playing in the allowance trading markets—generating friction that is designed to spur on action at other levels, and smoothing the friction by demonstrating how a response to climate change can be achieved. In the allowance trading markets, experiments are mainly playing this role in North America, where subnational and transnational emissions trading systems have simultaneously emerged into the void left by a lack of federal policy in the United States and Canada.[49]

The story of the three subnational cap and trade systems should be familiar by now—reacting to the recalcitrance of federal governments, subnational actors (both U.S. states and Canadian provinces) looked to move forward with their own response to climate change regionally by banding together in the East (RGGI), Midwest (Midwestern Greenhouse Gas Reduction Accord), and West (WCI). As the remarks (quoted above) of the governors from the U.S. West demonstrate, subnational responses, especially transnational ones, were ways of demonstrating leadership and pointing to the kind of response that could be generated more broadly if federal responses followed the lead of the states and provinces.

A partnership among 10 U.S. states (Connecticut, Delaware, Maine, Maryland, Massachusetts, New Hampshire, New Jersey, New York, Rhode Island, and Vermont), RGGI is the only one of the three subnational emissions trading systems that has gone operational at this point. It is a limited cap and trade system that puts a cap on emissions from the electrical generation sector and has the modest goal of reducing emissions 10% from this sector by 2018. Yet RGGI is being cautiously touted as a success. Trading began in 2009, and a key feature of RGGI has been its use of auctioning to distribute permits from the very beginning. Even with low prices (around $2 per ton), auctioning of permits have raised over $580 million from September 2008 to March 2009, much of which is recycled back into the states' economies in the form of grants for energy efficiency projects and building retrofits.[50]

The WCI and Midwestern Greenhouse Gas Reduction Accord cap and trade systems are still being finalized. The WCI is set to come online in 2012 with 9 participating states and provinces (British Columbia, California, Manitoba, Montana, New Mexico, Ontario, Oregon, Quebec, and Washington).[51] It will be an economy-wide cap and trade system that seeks a 15% reduction from 2005 greenhouse gas emission levels by 2020. The Midwestern Accord's design was finalized in June 2009 and is now awaiting approval from the regions' governors and premiers (Iowa, Illinois, Kansas, Manitoba, Michigan, Minnesota, and Wisconsin)—approval that may be challenging to achieve in the aftermath of gubernatorial elections in 2010 that saw a shift to Republican governors across the region. It is also an economy-wide cap and trade system with a slightly more ambitious goal of 20% reduction from 2005 greenhouse gas emission levels by 2020. Both the WCI and Midwestern Accord propose to regulate all six greenhouse gases and include auctioning as a means of allocating at least some of the permits to regulated entities.

The RGGI, WCI, and Midwestern Accord started with mixed motivations. In all three, concerns were stated about the urgency of the problem with concomitant references to the cobenefits to be had by pursuing climate action. Then Governor Kathleen Sebelius of Kansas captured these multiple drivers succinctly in proclaiming, at the inauguration of the Midwestern Accord, "As Governor, it's important to me to ensure that Kansas has a stable and reliable source of electricity that fosters strong economic growth. But it's equally important to me to protect our environment and water supply for future generations."[52] The WCI justifies its cap and trade program in exactly the same way:

> Analyses conducted on the WCI design indicate that the region can mitigate the costs of reduction emissions and realize a cost savings through increased efficiencies and reduced fuel consumption. These savings come in addition to the benefits for the region from a cleaner environment and promoting the kinds of investment and innovation that will spur growth in new green technologies and build a strong green economy.[53]

Yet these regional initiatives were born in a context of federal inaction across North America, and perhaps more than any other set of experiments examined in this research, these subnational emissions trading systems display the tension between a desire to create friction that spurs action at the federal level and to create a new way of responding to climate change experimentally.

Even though the regional cap and trade programs are justified in environmental and economic terms, at least part of the motivation for pursuing these experiments has been to spur on national action by creating a patchwork system that stakeholders (especially large industry) abhor. Recall Governor Richardson's comment at the inauguration of the WCI: "with this agreement, states are once again taking the lead on combating global climate change—while Washington, D.C. sits on its hands. . . . This historic agreement signals our commitment to tackling the problem head-on at the regional level and building on efforts in our individual states."[54] Franz Litz, one of RGGI's designers, recalled that when it was initially proposed by Governor George Pataki of New York, the response from some governors was tepid precisely because they thought that a federal system was desirable. Subsequently, at least one of the goals behind developing this regional approach was to make stakeholders want a federal system.[55] This is classic in environmental policy-making in the U.S. federal system. Innovative states move quickly and push the envelope, creating the friction that eventually drives the federal government to make uniform rules.[56] As Litz commented, these initiatives were designed precisely to create ripple effects, both within the regions pursuing them and in the broader North American context.

And they have created friction. I cannot claim that the federal cap and trade proposals passed by the U.S. House of Representatives in 2009 and considered, but rejected, by the U.S. Senate in 2010 are a direct response to the development of regional trading systems.[57] A number of factors (the 2008 election ushering in a president and Congress more amenable to action on climate change being perhaps the most important) drove the U.S. Congress to seriously consider a form of emissions trading in the 2009–2010. Yet the regional systems are pushing the discussion forward about whether and how to tackle climate change in North America. They are also generating questions about how to reconcile traditional national-level climate policy with experimental governance initiatives.[58] Indirect evidence of how regional trading systems have created national friction can be found in the Kerry-Lieberman legislation tabled in the Senate in May 2010. That legislation failed to become law, but it does provide insight into how the Congress viewed the regional trading systems. While the regional systems did demonstrate how cap and trade could be designed and elements from the regional systems were included, the Kerry-Lieberman bill unequivocally forbade states from enacting cap and trade systems.[59] This was the intended ripple effect—a federal response in reaction to a patchwork set of regional systems.

The federal-regional debate over cap and trade is a key concern because over time in the absence of federal policy, the regional systems began designing systems that would work in their particular circumstances and formed somewhat of a regional movement. They learned from one another. RGGI has been a touchstone in North America, with both the WCI and Midwestern Greeenhouse Gas Reduction Accord design processes taking lessons from the RGGI experience.[60] Both WCI and the Midwestern Accord call for economy-wide cap and trade. Patrick Hogan from the Pew Center on Global Climate Change stresses that the "RGGI experience was a good learning platform" and that without RGGI, the WCI and Midwestern Accord "would probably not have been as ambitious in scope."[61] The regional systems are also in discussions about linking. Litz reports that the three have now had two meetings where they have organized working groups on linking systems, offset policies, and low carbon fuel standards.[62] A 2010 white paper cemented this relationship, laying out the common offset policy developed in concert by the three regional trading systems and confirming the dual roles these systems are designed to play:

> The whitepaper is intended to serve as both an internal policy document for use among the programs and as a public policy document to inform the development of comprehensive climate policy in North America. As an internal document, the whitepaper articulates key quality requirements for offsets and offset programs to facilitate potential future linking of regional cap-and-trade programs. Future linking of programs could include coordination of offset programs and offset reciprocity among programs, which would require that each program maintain minimum offset quality requirements and standards. As an external document, the whitepaper communicates common underlying offset quality concepts that are incorporated into the design and implementation of each of the regional cap-and-trade programs to inform the design and implementation of national cap-and-trade programs in the U.S. and Canada.[63]

While the regional systems may have been designed as a backstop for failed federal action and as a way to be relevant in federal discussions,[64] they are actually designing systems to carry out cap and trade. This has simultaneously created a smoothing effect, in that these regional systems demonstrate how emissions trading systems can be designed and operated, and additional friction, as questions of how regional systems fit with a potential U.S. federal system, should one arise. Will a federal program preempt regional systems? Are they easy to incorporate? Can regional systems persist in the face of ongoing federal reluctance to move? These are open questions. The World Resources Institute has published research showing how incorporating regional systems into a federal cap and trade would be relatively simple.[65] They have also produced a report for how US climate policy could move forward in the absence of a federal cap and trade system, through state action.[66] Robert Stavins has recently argued that federal preemption of regional systems is imperative. The Kerry-Lieberman legislation proposed in the U.S. Senate demanded federal preeminence, but states pushed back against that position even before the legislation failed to pass, with some claiming that the federal legislation would not have been stringent enough and could have rolled back gains made regionally:

> Groups of state environmental chiefs, attorneys general, and US senators wrote the drafters of the federal bill in recent weeks, expressing concern that it could undo gains made under the Regional Greenhouse Gas Initiative.[67]

Given the uncertainty of federal legislation, more attention is now on the state regional systems as well.[68] They may provide the only means of actually meeting U.S. international commitments under the Copenhagen Accord should the current debate over climate in the U.S. Congress go unresolved or resolved negatively as the 2010 midterm elections might imply. At the very least, these processes are inextricably intertwined.

EXPERIMENTING IN THE CREDIT MARKETS

Starting from scratch is given a new meaning in carbon credit markets, in that they do not even begin with a tangible commodity. What has evolved is a set of interactions, relationships, and standards oriented toward monetizing promises to not emit and to reduce greenhouse gas emissions. The roles of climate governance experiments in the development and functioning of the credit markets outlined in figure 6.1 have been explicit and varied. Experiments that initiate emissions trading, like the CCX, serve as a source of demand for carbon credits (over 2 million metric tons of carbon dioxide equivalent credits were bilaterally traded just from September to December 2009 on the CCX).[69] The Climate Registry provides the infrastructure for measuring the carbon footprints that complements demand sources to produce actual consumption. Yet it is clear that climate governance experiments have been most prominent in a particular role in the voluntary credit market—setting standards for offset credits. The entire voluntary credit market is essentially an experiment spurred on by opportunities and perceived voids in the multilateral process and the carbon markets being developed within it. Climate governance experiments are a keystone in this market, providing the methods and information that create the very commodity being exchanged. Experiments are clearly not alone in providing this infrastructural service—experimental standard setters draw on the CDM experience and more generalized standard setting guidelines provided by the International Standards Organization[70]—but they have become a focal point around which the market is emerging.

Three prominent standard setting experiments are the Voluntary Carbon Standard, the Climate Action Reserve, and the American Carbon Registry.[71] Each fills a similar functional role in the processes shown in figure 6.1—providing a stamp of approval on offset projects and serial numbers for carbon credits, which turn the projects (efforts to reduce greenhouse gas emissions) into a source of commodities (specific credits in terms of metric tons of avoided or reduced emissions). These three experiments have each evolved slightly different approaches to standard setting—similar functional roles have produced diverse standard setting procedures and methods.

The Voluntary Carbon Standard, itself the product of interexperiment collaboration—between the Climate Group and the World Business Council on Sustainable Development—provides a broad set of principles to ensure the financial and environmental integrity of offsets and then approves project methodologies that are developed in the field. This bottom-up approach relies on validators— consultants who have been approved by the Voluntary Carbon Standard to validate projects on their standards—and the Voluntary Carbon Standard requires that two independent validators certify that a project is up to snuff. The process has four stages.[72] First, a project proponent proposes a new methodology for reducing emissions and creating offset credits and submits the proposal to the Voluntary Carbon Standard. Second, the proposal is made available for public comment and consultation. Third, the proponent of the new method contracts a validator to assess the method, and the Voluntary Carbon Standard contracts the second validator. Finally, the Voluntary Carbon Standard reviews the assessments and approves or denies the new methodology. Once approved, the project is eligible to generate offset credits—Voluntary Carbon Units—after the actual reductions have been verified by a third party.

The Climate Action Reserve, in contrast, works from a more top-down approach. This experiment develops methodologies for measuring and tracking offset credits in-house. They have developed their own protocols for a range of possible offsetting activities, including emissions and reductions associated with coal mine methane, forests, landfills, and livestock.[73] This top-down approach requires more work up front by the Climate Action Reserve in developing the protocols than is required in the bottom-up approach adopted by the Voluntary Carbon Standard, but the registering of projects at the back end is a simpler process:

> The Climate Action Reserve staff pre-screen projects for eligibility. Eligible projects are posted on the Reserve site with a status of "Listed." The

next step is verification by an independent, accredited verification body. Once completed, the Climate Action Reserve staff review the verification documentation, and if the project passes this final review process, it is labeled "Registered" and CRTs [climate reserve tons] are issued.[74]

The American Carbon Registry offers a third alternative for offset standard setting that utilizes both top down and bottom up mechanisms. They develop a number of methodologies in house as they employ a number of experts in the field—they have even developed methodologies for the VCS standard—but they also approve methodologies proposed by project developers.[75] The American Carbon Registry adds the element of scientific peer review to their standard setting process and they claim that their process allows for the quick and relatively inexpensive approval of high quality standards.[76]

Experiments are performing functionally similar roles but in different ways. In some sense, this is exactly what is necessary in a developing market because there were no standards as the voluntary credit markets emerged, and other than the CDM, which was roundly criticized, there were few models of how to set standards. From a self-organization perspective, this kind of redundancy in the system is expected as new types of behavior are experimented with. Mary Grady reflected that "everybody is focused on quality" but that the question was "how do you get there" efficiently.[77] Yet the initial burst of standard setting that has been apparent in the last couple of years is already starting to recede.[78] David Antonioli, CEO of the Voluntary Carbon Standard, notes that a similar level of rigor now underpins most standard setting endeavors, and that we are starting to see market consolidation (albeit slowly) around a few prominent standards.[79]

The standard setting experiments that are operating are not doing so in a vacuum. On the contrary, the major standard setters are linked both cooperatively and competitively. First, a relatively small community of people is engaged in this sector, and many have worked on more than one standard setting exercise. For example, Toby Janson-Smith, who previously directed the Climate, Community and Biodiversity Alliance, recently helped develop the agriculture and forestry program for the Voluntary Carbon Standard.[80] The Rainforest Alliance is officially a member of the Climate, Community and Biodiversity Alliance and the CCX, but also serves as an auditor for the Voluntary Carbon Standard, with whom they are "in touch on a weekly basis on interpretation" of protocols.[81] The web site for the American Carbon Registry touts its connections as follows:

> Team members have been participants in standards and methodologies manuals and publications and co-authored protocols for the Intergovernmental Panel on Climate Change (IPCC), Clean Development Mechanism (CDM), U.S. Department of Energy (DOE) 1605(b) program, U.S.

Environmental Protection Agency (EPA), U.S. Department of Agriculture (USDA), the USDA Forestry Service, the World Bank, the International Tropical Timber Organization, UN organizations, the Voluntary Carbon Standard Association (VCS) and the Climate, Community and Biodiversity Alliance (CCBA), among others.[82]

Beyond personal connections, the standard setting activities and organizations are linked functionally. The standards being set are not mutually exclusive, and there is a good deal of cooperation and mutual recognition among the standard setters. For instance, the Voluntary Carbon Standard recognizes protocols from the Climate Action Reserve:

> Thus all of the Reserve's project protocols are approved methodologies under the VCS [Voluntary Carbon Standard]. This means that all reductions verified under Reserve protocols and registered as CRTs [climate reserve tons] on the Reserve . . . are eligible to be converted to Voluntary Carbon Units (VCUs).[83]

Similarly, the American Carbon Registry announced in January the purchase of 100,000 metric tons worth of credits by Entergy that were originally listed as voluntary carbon units with Voluntary Carbon Standard:

> Originally certified by the Voluntary Carbon Standard, which is known for its high quality offset criteria, and listed on Markit Environmental Registry, the offsets were reverified to the American Carbon Registry Standard, delisted from Markit and reissued on the American Carbon Registry. The transaction marks the first time in North America where multiple carbon registry programs have worked together to manage offset project accounting.[84]

Cooperation is not the only kind of interaction, however. Given their market orientation, even though these three experiments (the Climate Action Reserve, the Voluntary Carbon Standard, and the American Carbon Registry) are nonprofit organizations, there is still competition. Again, like the experimental interactions seen in chapter 5's analysis of technological deployment in cities, this competition is the not necessarily the traditional zero-sum variety. Derik Broekhoff, vice president of policy for the Climate Action Reserve, notes that standard setters are not actively looking to steal market share, but that they are competitors in some sense.[85] There are numerous zones of agreement, not least the underlying philosophy that offsets can be done and done well in the service of responding to climate change and developing carbon markets.[86] Competition is over the development of quality standards and efficient processes and finding niches in what could potentially be a very large market. In fact, all the staff involved with standard setting experiments with whom I talked in the course of this research talked about competition among standard setters as a positive dynamic.[87] Different standard setters provide a range of options for developing projects and methods of offset production,[88] and this allows for "testing out different models and serving different segments of the market."[89] Mary Grady explicitly argued that competition is good for the sector because it generates innovation in both project development (i.e., methodologies for producing carbon reductions) and standards (ways of verifying reductions). Innovation and lots of activity, she says, is what will be necessary to scale up to "where we need to be."[90]

At this point, competition is as much about finding niches in the realm of offset standard setting as it is about direct rivalry among experiments, though it should be clear that not all standard setters are likely to continue to exist as the market evolves and develops. The experience of the Climate, Community, and Biodiversity Alliance is instructive in this sense. This experiment emerged, like the others already discussed, because of concerns about standards in the voluntary offset market. But, rather than contributing an additional set of standards for measuring and verifying emissions reductions, this experiment focuses on supplementary standards to ensure that forest offset projects have a positive impact on local communities and biodiversity. Joanna Durbin, director of the Alliance, recounts that when this experiment emerged, while forest carbon projects were considered a good idea, there was "lots of heat because there were some bad forest projects not doing what they said they were doing."[91] The CEO of BP at the time (Lord John Brown), who was on the board of Conservation International (an initiator of the Alliance), according to Durbin, challenged the NGOs to define what was a good project.[92] The Climate, Community, and Biodiversity Alliance emerged to meet this challenge and launched a standard in 2005 (revised in 2008) that was designed to be pursued in conjunction with a greenhouse gas emissions standard—in other words, project producers could get Climate, Community, and Biodiversity Alliance certification on top of Voluntary Carbon Standard approval, for instance. Now Durbin claims that "the majority of forest carbon project aspires to CCB standards," a claim that is bolstered by survey results from offset purchasers.[93] Over 100 projects are now seeking their approval.[94]

As noted, the voluntary credit market is still relatively small, but the impact of standard setters goes beyond the voluntary credit markets, and experimentation in this sector has the potential for enormous ripple effects. In effect, standard setters are supplying key infrastructure for the whole global carbon market. In fact, we are witnessing a blurring of the distinction between the experimental and traditional governance systems. Offset credits are playing an increasingly prominent role in the design of emissions trading venues, especially ones set up and managed by nation-states.[95] Recent proposals for a U.S. federal emissions trading system contained provisions for 2 billion tons of offset credits per year.[96] Experimental standard setters are vying to be included in these regulated systems and are playing a role in shaping how offset credits will be used in emissions trading venues. Antonioli is heartened to see the competitive nature of the standard setting in the voluntary markets because he is convinced that this kind of innovation provides frameworks that compliance markets (traditional, like the EU emissions trading system, and experimental, like the subnational approaches in North America) can pick up.[97] Governments, corporations, and other participants in carbon markets are learning lessons in the voluntary offset world that can feed into regulated systems.[98]

Conclusion

Since 1992 when Richard Sandor foresaw the need to build institutions to bring forth a new commodity market based on greenhouse gas emissions, carbon markets have enjoyed a meteoric rise. The bulk of the global response to climate change is market oriented, and the center of gravity of the global response, even with the recent stumbles on the part of the United States and Australia,[99] is bound up with market mechanisms—credit and allowance markets. In this chapter, in addition to giving an overview as to how these markets have and are developing, I have endeavored to illustrate the key role climate governance experiments are playing in this broadly experimental activity. A division of labor does seem to be emerging, with the development of specific roles and niches in the carbon markets. The competitive and cooperative interactions are organizing seemingly disparate initiatives into a coherent cluster of activity.

However, in contrast to the cluster of activity discussed in chapter 5, there is more palpable interdependence between the experimental governance system and the traditional governance system in the area of carbon markets. The carbon markets themselves have become a site of governance—a site where multiple efforts converge in the attempt to govern climate change.[100] The multilateral process catalyzed the emergence of carbon markets, but they quickly moved beyond the control of the UN treaty-making process that spawned them, growing geographically and functionally and fragmenting into multiple markets. Climate governance experimentation in this area thus goes beyond merely reacting to perceived voids in traditional governance systems. On the contrary, standard setting experiments in the credit markets have created a niche and now play a key role in creating the very commodity the market is built on. Regional emissions trading systems are the focal point of carbon markets in North America. These efforts shape how the U.S. and multilateral systems will engage with carbon markets, while in turn, decisions emanating from the traditional governance system in part determine the fate of these experiments.

The ultimate potential of carbon markets to lead the way toward decarbonization is still uncertain. The global financial crisis has sapped necessary capital for pursuing offset projects, depressed the demand for offset credits,[101] and increased governments' and publics' wariness about the whole nature of carbon markets. Clearly, a number of actors are willing to pursue and the development of carbon markets, both allowance and credit, and have interests in doing so. At the end of the day, however, what this chapter demonstrates is that neither experiments nor traditional governance practices have any meaningful existence without the other in this arena. Experimental initiatives need national and global policy if they are to be effective in building the infrastructure of carbon markets. National carbon markets and efforts at creating a linked global carbon market will rely on experimental initiatives that innovate and that have experience designing and navigating emissions trading and offset systems. Carbon markets cannot function solely in the experimental system or in the traditional multilateral system. Leadership is required to ensure that they work together effectively.

7

Lost in the Void or Filling the Void?

We have a thing called the "cheerful disclaimer"—which
means we have no idea if the idea is going to
work or not. It's an invitation to have a go.
—Rob Hopkins, founder of Transition Towns (2010)

[Y]ou're not a complete idiot. You know there is a big
difference between finding islands of excellence and
creating a national system based on them.
—David Brooks, columnist, on U.S. health care reform
(July 24, 2009)

[C]limate change is not a discrete problem amenable to any
single shot solution, be it Kyoto or any other. Climate change is
the result of a particular development path and its globally
interlaced supply system of fossil energy. No single intervention can change such a complex nexus. . . . There
is no simple silver bullet.
—Gywn Prins and Steve Rayner (2007)

A Tale of Two Copenhagens Reimagined

*During the 2009 Copenhagen climate change negotiations, activists strode through
the Bella Center festooned in eye-catching costumes, berated the "Fossils of the
Day" for obstructing progress on a global accord, and sought to inform and influence*

national delegations. Harried and exhausted negotiators diligently worked long hours on a series of specific, limited agreements that would serve to smooth the functioning and linkage of carbon markets, cement cooperation on the development of large scale pilot projects and technology transfer in a number of sectors, and facilitate the transnational activities of subnational governments. Side events focused on the linkages between city initiatives and global carbon markets, voluntary standards and compliance verifications in the credit markets, linking subnational and national allowance markets, carbon accounting, adaptation to climate change, and means of bringing international financial institutions into the deployment of climate-friendly technology. In the corridors, academics and students huddled in small groups trying to make sense of the range of trajectories being discussed and the obstacles to and means of creating synergies in what has become a more encompassing global response to climate change. Cameras and microphones were ubiquitous as the media sought to convey the complexity of the global response to climate change. Logistical and security challenges unfolded as heads of state swooped in during the final days of the conference to sign a range of agreements and recommit to an overarching framework goal. Outside the Center itself, demonstrators chastised the negotiators and urged them to go faster, attend to equity concerns and adaptation to climate change, and beware an overreliance on market mechanisms. The eyes of the world focused on the events at the Bella Center. Many are now more hopeful at what they witnessed.

Fanciful? Certainly—especially given the political context within which I researched and wrote this book. The misplaced optimism with which many went to the Copenhagen negotiations in December 2009 has been replaced in 2010 with downright despair. The multilateral process looks as hopeless as ever with the addition of the Copenhagen Accord to already contentious negotiations over the Kyoto Protocol replacement. The deadlocked U.S. Congress appears to be as far from acting on climate change as it was during the Bush Administration. The global financial crisis of 2007–2009 has sapped resources for pursuing significant action, and the promised "green" stimulus programs have not been as revolutionary as hoped. Climate science has come under renewed attack from skeptics emboldened by missteps of the Intergovernmental Panel on Climate Change. A disastrous oil spill in the Gulf of Mexico has highlighted both the urgency of moving beyond fossil fuels and the distance the world needs to go politically and technologically to make this move.

This book is not a full antidote to the recent doldrums surrounding the global response to climate change, but it is a call to think differently about climate governance in ways that might provide a bit more grounded optimism. I am not advocating wishing the world were different. On the contrary, through the course of this book I hope to have demonstrated that we need to think differently about climate governance precisely because the global response to climate change is unfolding differently from what we have come to expect. We need to change our perspective, both academically and practically, in order to understand, take advantage of, and further the changes in climate governance that are already taking place.[1]

Taking Experimentation Seriously

The analysis in this book is a beginning rather than a definitive account; the experiments chronicled here have only recently gained momentum, and their fates have yet to be conclusively determined.[2] Even so, this book has accomplished a number of tasks that collectively suggest new ways of thinking about climate governance and the global response to climate change. First, it has simply revealed the variety and extent of activity being undertaken beyond or only loosely connected to the multilateral negotiations. Even with a severely restricted definition of a climate governance experiment, I have found almost 60 innovative initiatives that are operating outside the Kyoto process at all levels of political organization. There is significant momentum in the global response to climate change that is not captured when we concentrate our attention on multilateral treaty-making.[3] And to be frank, this study only captures of a fraction of what is occurring. An enormous amount of activity is going on around climate change both in the experimental system I have defined and outside that system. Mike Hulme is substantially correct in his argument that climate change is a defining feature of modern life.[4]

Second, experiments are organized. I have demonstrated that climate governance experiments, rather than being disparate or random, are both patterned and explicable. Experimental governance activity is developing in a relatively coherent fashion. Diverse climate governance experiments initiated and implemented by the full range of political actors share a philosophical underpinning and worldview of liberal environmentalism.[5] The experiments are cut from the similar cloth. Yet this shared ethos is not the only defining feature of the collection of experiments. They also exhibit both functional differentiation and thematic clustering. Experiments can be readily grouped into four archetypes (networkers, infrastructure builders, voluntary actors, accountable actors) on the basis of how they are responding to climate change. In addition, basins of attraction or clusters of activity have emerged in a number of areas facilitated by this functional differentiation, including technology deployment in cities and carbon markets—sites where experiments interact and build relationships with each other as well as with other participants in the global response to climate change. Differentiation and clustering are organizing a nascent experimental system that may be developing into a division of labor and is certainly generating friction with traditional governance mechanisms. Experiments are creating and filling niches and exploring a range of means of responding to climate change.

Third, I have provided an analytically grounded story that makes sense of the emergence and patterns of experimentation. The key dynamic is feedback between actors and the governance context and the way this feedback produces centripetal or centrifugal processes. Climate governance experiments are both symptoms and drivers of growing centrifugal forces in climate governance. Their emergence was made possible by growing uncertainty in the governance context about the legitimacy and effectiveness of multilateral treaty-making for climate change, or at least concerns about relying entirely on this mechanism to drive the global response. Experimental momentum has grown in the last decade, further enhancing the uncertainty around multilateral governance. The patterns apparent in the population of experiments is evidence that climate governance has gone beyond producing novelty and is in the process of normalizing into a new governance context, which is characterized by interactions between the traditional multilateral system and the increasingly coherent experimental system.

Fourth, my examination of experiments and their activities in chapters 4–6 has revealed how the experimental system is organizing functionally and thematically, lending plausibility to the analytic account developed in chapter 3. Experiment initiators did react to the broad sources of uncertainty like the signing of and stalemate that followed the Kyoto Protocol. Experiments are self-organizing—interacting, cooperating, competing, and filling niches in functionally specific ways and in specific sites of climate activity. They are in the process of transforming novelty into normalcy and building an experimental system that has some coherence and interacts in important ways with the traditional multilateral governance system. Questions remain as to whether the experimental system will develop into a friction and synergy-inducing division of labor among experiments (and beyond) or a sorting process whereby collective aspirations of addressing climate change are hamstrung by actors' interests and freedom to experiment.

Finally the descriptions of experimentation throughout the book suggest that the experimental system provides a source of friction that can motivate other actors to take climate action and a way to smooth friction by providing the institutional, political, and technological capacity to act quickly. This remains a suggestion but, hopefully, an intriguing one. The climate governance experiments I have examined all *intend* to engender something in the way of friction and smoothing. I have provided evidence that they are carrying this out. Yet additional time, results, and research are necessary before a definitive conclusion can be reached about the role of experiments in driving forward the global response to climate change.

Thus, the preceding analysis confirms that the center of gravity in the global response to climate change is shifting, but by no means can I claim that the experimental system has solidified into a new orthodoxy. Novelty is still being normalized, and as I will discuss below, this provides a window of opportunity for leadership. Climate governance experiments can and are serving as catalysts to a whole range of changes—technological, institutional, economic, and political. There is significant momentum in the experimental system and a flourishing creativity that needs to be widely unleashed. The potential is there for significant transformation and disruptive change; it just has yet to be fully realized. The experimental system is not ready to replace the multilateral system, and perhaps it never will, but I am convinced by the evidence presented here that climate governance experiments will grow more prominent in the near future and are poised to be the driving force behind the global response to climate change.

In light of these conclusions, my hope is that this book enables readers to think differently about climate governance and the global response to climate change. At one level, it should be taken as a source of realistic optimism at a time when the prospects for an effective global response to climate change appear grim. Gywn Prins and Steve Rayner, whose quotation opens this chapter, make a strong case about the need for "silver buckshot" instead of a "silver bullet" in the response to climate change.[6] The preceding chapters have chronicled the current content of that buckshot. Numerous significant actors are working toward transformation of the energy system, economy, and society. The continued failure of the multilateral approach to produce a top-down "solution" to climate change is disappointing but should not be seen as fatal. In fact, given the nature of climate change and multilateral negotiations, as well as the enormity of the task at hand, we cannot expect a single multilateral treaty to structure the global response.[7] Such an expectation too often blinds us to the momentum in other processes and places—like the experiments examined in this research—that both are moving more quickly than the multilateral process and may be better suited for the problem.

This book thus lends further credence to recent calls for rethinking climate governance and for better theorization of multilevel dynamics.[8] There is no shortage of new and exciting ways to think about how climate governance could be understood and improved. Yet many remain frustratingly wedded to a focus on nation-states and multilateral treaty-making.[9] Megamultilateral treaty-making is, as Prins and Rayner have called it, "the wrong trousers," a serious misfit, for the global response to climate change.[10] In examining experimentation and providing a way to make sense of it, this book provides a way to move beyond state-centrism without having to make the indefensible claim that multilateral treaty-making and nation-states are no longer relevant. What this analysis instead implies is that the dynamics in the experimental system may be as important as those in the traditional governance system and that experimentation can provide the impetus for the global response in a way that multilateral treaty-making simply cannot.

Further, silver buckshot may not be the correct metaphor either. This could imply atomized initiatives and projects working towards a response to climate change independently from one another. Reality in the experimental world is much different. Functional differentiation and self-organized clumpiness instead describes the structure of experimental initiatives. The connections, relationships, and sites of convergence among climate governance experiments is precisely why the global response to climate change looks different today and why there is hope that activity beyond the multilateral negotiations has the potential to catalyze disruptive change towards an effective response.

More research is necessary, however, before the relevance of climate governance experimentation can be fully confirmed. Three topics are especially pertinent. The first is the power and politics of experimentation.[11] This study has treated experiments as relative equals in an attempt to ascertain the broad contours of fragmenting climate governance. Now that the idea of an experimental system is established, the next step is to turn an analytic eye on the politics of and between experiments. If a relatively independent experimental system is actually emerging, the power dynamics between experiments—how resources are authoritatively allocated among experiments—will be a crucial area of study. This is not radical revision of academic inquiry; on the contrary, this is a call to bring to bear to traditional concerns (power, authority, legitimacy, and contestation) to a new political context. Taking experimentation seriously means treating climate governance experiments as political entities and examining the politics of their interactions within and beyond the experimental system.

The second area is research on effectiveness. Earlier chapters have hinted at the difficulty of assessing the effectiveness of experiments individually or in concert, but given that this book posits and strongly argues for the relevance of experimentation, the next step is to assess that argument not in terms of the development of governance mechanisms but in terms of addressing the problem of climate change. If the center of gravity is moving toward an ineffectual governance configuration, we need to know quickly.

But of course, evaluating effectiveness is a vexing problem, necessary but difficult to accomplish. The metrics that we use to evaluate the effectiveness of climate governance experiments are not clear. On the one hand, their ultimate worth is depends on achieving a stabilization of the climate system and a turn to a low carbon economy. Yet, tracing the causal impact of climate experiments on these outcomes (should they come to pass) is virtually impossible given the vast array of forces and dynamics that would contribute to such a transformation. Yet we can evaluate the impact and effectiveness of climate governance experiments more modestly, namely if and how they contribute to trajectories of change. Further, in addition to examining the effectiveness of individual experiments, this book makes the case that because initiatives participate (consciously or unconsciously) in an experimental system, the collective impact of climate governance experiments must be a focus as well.

This book analyzed how experiments, both in the abstract and in specific instances, individually and collectively, can and have served as a catalyst for change in the governance context. With that dynamic established the necessary work yet to come becomes clear—extending the analysis of friction and smoothing to systematically evaluate how experiments and experimentation contribute to trajectories of change. A few key scholarly works have begun to assess how and under what conditions initiatives like climate governance experiments can engender ripple effects that will significantly increase their impact on the global response to climate change. Steven Bernstein and Benjamin Cashore have posited means through which transnational nonstate actors can influence nation-states' policy-making.[12] They concentrate on specific levers available to transnational actors like opportunities in the global market, existing "international rules and regulations; changes in international normative discourse; and infiltration of the domestic policy-making process."[13] Henrik Selin and Stacy VanDeveer similarly identify four "pathways to policy change" for subnational actors hoping to influence national and global policy. They include "(1) the strategic use of demonstration effects, (2) market pricing and expansion, (3) policy diffusion and learning, and (4) norm creation and promulgation."[14] These are all essentially ways that climate governance experiments can create friction in and provide smoothing infrastructure for the global response to climate change, catalyzing action beyond the experimental system. While the analysis in this book has demonstrated that experiments are undertaking these activities, more research is necessary to systematically apprehend the friction and smoothing roles of climate governance experiments.

Finally, significant ethical analysis of climate governance experimentation will be crucial. As one example, chapter 2 reported on the liberal environmental foundation of climate governance experimentation, and chapter 6 discussed the development of carbon markets. Both chapters did so without fully engaging the ethical contestation around the market orientation of the global response to climate change. This does not imply that such contestation is irrelevant or muted. Significant protest over carbon markets characterized the 2009 Copenhagen negotiations, and depicting market orientation as the "compromise"[15] of liberal environmentalism is not complimentary. Thus, in addition to questions about the efficacy of climate governance experimentation, legitimacy concerns should also face critical scrutiny.[16]

I understand the draw of the megamultilateral process.[17] There is comfort in a single authoritative response that encompasses the globe and achieves binding treaties that can hold nation-states accountable for taking action on climate change. A proliferation of experimental governance responses leads to uncertainty on both effectiveness and legitimacy grounds. In part I called these initiatives experiments precisely because it is unclear if they will succeed or fail or even if they are good ideas. The calls for smaller, like-minded groups responding to climate change at multiple scales through a variety of tools might make sense.

There is, after all, plenty to do in responding to climate change. Some experiments will certainly fail and others will evolve (the 2010 cessation of trading in the CCX is a pertinent example). But experimentation and the self-organization of the experimental system will continue apace in the near term. Will the whole be at least as much as the sum of the parts, and will the sum of efforts be enough and be legitimate? If the age of effective megamultilateralism in global environmental governance is over (assuming it ever existed outside our imaginations), we have yet to reach the age of effective experimentation, and the legitimacy of multiple initiatives at multiple levels of political organization has yet to be established. This book has laid out a way to think about the dynamics of climate governance experimentation that brings these questions to the fore and reveals the *potential* in the experimental governance system. Additional study is necessary to ascertain whether and how the potential can be realized, but a reframing of leadership in climate governance is necessary to bring to fruition the potential that experiments have for transforming the global response to climate change.

Envisioning Leadership in an Experimental World

Uncertainty and instability present opportunities as well as challenges. If the Kyoto process is as ossified and ineffective as its critics claim, an uncertain response made up of climate governance experiments across the world can hardly do worse and may do better. But before it can reach its full potential, we need to redefine leadership in climate governance to account for the shift in the global response to climate change. The analysis developed in this book suggests that the global response to climate change has reached a critical juncture. The multilateral system is foundering, but the experimental system is not yet fully developed. Leadership on climate governance is needed and will prove most efficacious in the coming years at the boundaries and in the linkages between these systems. There are opportunities to influence how the novelty generated in the last decade becomes normalized and how the experimental system will interact with the more traditional governance mechanisms that have heretofore dominated the international response to climate change.

Leadership is needed to ensure that a division of labor wins out over Tiebout sorting. Chapters 4–6 raised the possibility that an efficient of division of labor is emerging in the experimental system of governance but could not rule out the possibility that Tiebout sorting is the more likely outcome of experimentation. A division of labor process or outcome can and must be fostered in the experimental governance system.

The first step is merely recognizing that there is an experimental *system* of governance that can provide resources for furthering momentum on climate action. During the course of the research for this book, on a number of occasions the description of the project given to informants elicited the response "we really need to know about this." I inferred that people working within the experimental system were not entirely aware of the extent and contours of the system itself. I asked almost every person working with climate governance experiments whether they felt like they were part of a broader movement. To a person, they all answered in the affirmative, yet describing the content or contours of the broader movement—of the experimental system—is not something those participating in climate governance experiments have spent much time or energy doing.[18] Some informants even asked me about what other organizations and actors were working in the sector. To be clear, and to reinforce a lesson from the previous chapters, experiments are linking and interacting. In fact, observing an experimental *system* is only possible because of the organization and interactions of experiments. However, this does not mean that the big picture is clearly in the focus of the experimenters. One intention for this research has been to shed some light on the big picture of climate governance experimentation precisely so that the networking that has already emerged between experiments can expand. Understanding how the experimental system is organized will facilitate creating a division of labor within that system.

Moving from envisioning a division of labor to building one requires enhancing the connections and niches that are already emerging. To their great credit, most of the experiment initiators examined in this research realize this and have made it a priority. Dasha Rettew of The Climate Group told me that moving ahead with making more connections is a key element of their strategy moving forward. They are looking for additional networks within which they can disseminate the results of their pilot studies to further the momentum behind market transformation.[19] More of these connections can be built, taking advantage of the common language that all these experiments speak—based on the premise that climate change can be addressed in a way that enhances economic growth[20]—and the functional differentiation and activity clusters in the experimental system.

One possibility would be to work on more closely linking the activity clusters discussed in chapters 5 and 6. One activity cluster is working through various programs to develop and deploy technology in municipal networks. A number of "globally relevant" pilot projects are being designed and implemented.[21] This cluster of activity is interesting but is consistently hamstrung by scarce financial resources. The other activity clusters are in the carbon markets. Here you have a number of initiatives aimed precisely at monetizing greenhouse gas emissions and reductions. The linkage between these clusters needs to be pursued. This has not gone unrecognized. One experiment—the Carbon Finance Capacity Building Program—is explicitly oriented toward the generation of carbon credits by undertaking projects in cities.[22] The program is designed to help cities turn projects on buildings, energy, and transportation into carbon certificates from the achieved emission reductions using the project-based mechanism of Carbon Finance such as the Clean Development Mechanism (CDM) or voluntary certification schemes like the Gold Standard (GS) of Voluntary Carbon Standard (VCS).[23]

Their programs are limited with a primary focus on 4–5 megacities in developing countries. Yet the goal is to form

a South-North-mega-city climate alliance network market, that allows emerging mega cities of the South to sell the generated certificates to an industrialized partnering mega city (e.g., within the C40 cities network, which is a CFCB [Carbon Finance Capacity Building Programme] partner).[24]

This could be a model for linking carbon markets and municipal technology deployment programs.

Unfortunately, too often a firewall seems to stand between these activity clusters and between levels of political organization. For instance, municipal climate actions are simply not on the radar screen for emissions trading initiatives in North America.[25] One informant characterized municipal action and regional trading systems as complementary but separate paths.[26] This may be changing. The agenda for the 2010 Carbon Expo—a key annual conference focused on carbon markets—had an entire lineup of panels dedicated to the topic "Linking Climate Finance and Carbon Finance." One panel focused exclusively on financing municipal climate action, in part from carbon markets, and included speakers from both the Clinton Climate Initiative and The Climate Group.[27]

Fostering a division of labor in the experimental system will involve bringing experiments that are currently on separate paths together in ways that take advantage of the different strengths and models that experiments possess and work through. The alternative is the kind of Tiebout sorting discussed in chapter 3, where initiating and implementing actors create and work within experiments that fit their own needs and interests with little interaction or interdependence. One key aspect of leadership in climate governance will be a matter of fostering the division of labor and avoiding sorting.

The second area for leadership is found at the boundary between the experimental and traditional systems of governance. The two systems of governance—traditional multilateral and experimental—have never been entirely independent. Climate governance experiments emerged in a context dominated by the existing multilateral governance system. They have been conditioned by that context and have been interacting (to varying degrees) with the traditional governance system since the very inception of climate governance experimentation. I am not predicting the imminent demise of multilateral climate governance. The system of international treaty-making is a constitutive feature of the international system, and states will continue to strive for a global treaty for the foreseeable future.[28] In addition, the annual COPs are about more than negotiations. They serve as a focal point for all kinds of actors (NGOs, corporations, students, interested individuals, and media), enhancing the spotlight on this key problem and communicating the sense of urgency that surrounds it.[29] No one involved in experimentation with whom I have talked claims that experiments can ultimately solve the climate change problem in the absence of an effective global treaty.

But two decades of failed treaty-making and evidence of an emergent experimental system convince me that the role of multilateral treaty-making in the response to climate change is destined to change. Rather than, or in addition to, negotiating comprehensive treaties designed to structure the overall global response to climate change, there will be an increasing need for smaller, more targeted negotiations. The friction that is emerging from the experimental system will provide ample opportunities to cement multilateral cooperation on *specific* issues. Multilateral climate negotiations will not and should not be abandoned. Yet we may need to adjust our thinking on what they can accomplish and the kind of issues they are best suited for addressing.[30]

Some analysts are calling for such a "Back to the Future" style of smaller negotiations more reminiscent of the early ozone depletion negotiations.[31] Richard Benedict, chief U.S. negotiator during the ozone depletion negotiations in the 1980s, argues that

> the climate problem could be disaggregated into smaller, more manageable components with fewer participants—in effect, a search for partial solutions rather than a comprehensive global model. An architecture of parallel regimes, involving varying combinations of national and local governments, industry, and civil society on different themes, could reinvigorate the climate negotiations by acknowledging the diverse interests and by expanding the scope of possible solutions.[32]

But calling for smaller negotiations in and of themselves still makes treaty-making the driving force of the global response to climate change. On the basis of the findings in this book, I contend that treaty-making must instead be used to ratify and further developments in the experimental governance system. As climate governance experiments create friction and offer smoothing solutions, limited scope multilateral treaty-making can be used to facilitate scaling up, linking, and furthering the dynamics bubbling up from below. The driving force should be coming from the experimental system of governance, and multilateral treaty-making can be used to enhance the innovations found there.

For instance, as national and even subnational emissions trading systems come online, a host of international cooperation issues will arise. Already at the

Copenhagen negotiations, the side events sponsored by the International Emissions Trading Association focused prominently on these issues of linking systems and cobbling together a global carbon market from disparate allowance and credit systems. Participants were especially interested in the monitoring, reporting, and verification concerns involved in ensuring the fungibility of carbon credits. Other discussions centered on how to deal with the economic pitfalls of diverse cap and trade systems emerging—capital flight to areas that fail to regulate carbon being among the most urgent—and the border adjustment or trade measures that would be needed to ameliorate them.

Multilateral negotiations and actions could also serve to enhance the transnational municipal actions discussed in chapter 5. The World Bank is already involved, peripherally, as a partner in the Carbon Finance Capacity Building program, but more could be done. As "globally relevant" pilot projects ramp up, there is a role for traditional governance mechanisms like international treaty-making and international organization facilitation to enhance and scale up technology deployment. Phil Jessup made this abundantly clear when he described the obstacles to implementing the global LED pilot project—there is a role for multilateral processes to facilitate the navigation of nation-state boundaries that transnational initiatives of climate governance experiments must undertake.[33]

Leadership from within the experimental governance system is therefore about pushing as hard as you can in every area that you can, developing innovations and relationships. Generate friction that creates interests and constituencies necessary for action in other areas and other levels. Generate the smoothing infrastructure that provides ready-made examples of how climate action can take place. Find niches and link initiatives. Leadership from the traditional multilateral governance system is a matter of restraining ambitions for a comprehensive global treaty and instead using treaty-making tools to enhance the burgeoning activities already taking place—a global framework signaling that climate will be addressed seriously along with agreements to regulate specific activities. In the search for innovation, "creating space for agents to build up alternative regimes is crucial."[34] Leaders in both systems need to recognize the boundaries and become comfortable working in the interstices between them.

Conclusion

To quote the underappreciated philosopher Ty Webb, "the shortest distance between two points is a straight line. . . . in the opposite direction."[35] In almost any system (governance, technological, social, economic), when innovations emerge, they "may appear as wholly inappropriate for a system."[36] Addressing climate change through the experimental system described in this book may indeed seem far-fetched and even a bit naive.[37] I am convinced otherwise. The governance experiments in the global response to climate change are new, but even at this stage we can see hope for innovative change. None of the experiments exists in isolation. There are clear and growing linkages and clusters of activities. They are pushing the envelope of climate action and demonstrating what is possible. These efforts provide hope that the world will respond effectively to the climate crisis. Further study, work, and leadership are required to see the potential of climate governance experiments fulfilled—to see them fill the void in climate governance and contribute to an effective and legitimate global response to climate change.

Appendix

Table A.1 **Experiment Web sites**

Experiment	Web site(s)
2degrees	About - http://www.2degreesnetwork.com/about_us/ what_is_2degrees/ FAQ - http://www.2degreesnetwork.com/faqs/
Alliance for Resilient Cities	http://www.cleanairpartnership.org/files/2009%20ARC%20 Membership%20Package.pdf
American Carbon Registry	FAQ - http://www.americancarbonregistry.org/aboutus/ faqs
American College & University Presidents Climate Commitment (ACUPCC)	About - http://www.presidentsclimatecommitment.org/ html/about.php The Commitment - http://www.presidentsclimate commitment.org/html/commitment.php Backgrounder - http://www.presidentsclimatecommitment. org/pdf/climate_leadership.pdf
Asia Pacific Partnership on Clean Development and Climate (APP)	Overview - http://www.asiapacificpartnership.org/About. htm FAQ - http://www.asiapacificpartnership.org/APPFAQ.htm APP Charter - http://www.asiapacificpartnership.org/ Charter.pdf
Australia's Bilateral Climate Change Partnerships	Main site - http://www.climatechange.gov.au/international/ partnerships/index.htmll
Business Council on Climate Change (BC3)	About BC3 - https://www.bc3sfbay.org/page/48 Principles - https://www.bc3sfbay.org/page/42 FAQs - https://www.bc3sfbay.org/page/43

continued

Table A.1 (continued)

Experiment	Web site(s)
C40 Cities Climate Leadership Group	2005 London Summit: http://www.c40cities.org/docs/summit2005/ 2007 Communiqué: http://www.c40cities.org/docs/communique_2007.pdf 2009 Seoul Summit: http://www.c40seoulsummit.com/
California Climate Action Registry (CCAR)	About: http://www.climateregistry.org/about.html
Carbon Disclosure Project	CDP questionnaire - http://www.cdproject.net/questionnaire.asp FAQ - http://www.cdproject.net/faqs.asp
Carbon Finance Capacity Building Programme	About - http://www.lowcarboncities.info/about-the-programme.html Mission - http://www.lowcarboncities.info/about-us/vision-mission.html
Carbon Rationing Action Groups	About CRAGs - http://www.carbonrationing.org.uk/wiki/about-crags? CRAGS - A Short Guide - http://www.carbonrationing.org.uk/wiki/crags-a-short-guide?
Carbon Sequestration Leadership Forum	http://www.cslforum.org/index.html?cid=nav_index
CarbonFix	The Concept - http://www.carbonfix.info/CarbonFix_Standard/Concept.html About - http://www.carbonfix.info/CarbonFix/Association.html
Chicago Climate Exchange (CCX)	Overview - http://www.chicagoclimateexchange.com/content.jsf?id=821 History - http://www.chicagoclimateexchange.com/content.jsf?id=1 FAQ - http://www.chicagoclimateexchange.com/content.jsf?id=74
Climate Alliance of European Cities with Indigeouns Rainforest Peoples (Klima-Bündnis / Alianza del Clima e.V.)	Mission and Organization Statement: http://www.klimabuendnis.org/our-objectives.html

Experiment	*Web site(s)*
Climate, Community, and Biodiversity Alliance (CCBA)	Mission - http://www.climate-standards.org/mission/index. html Fact sheet - http://www.climate-standards.org/ pdf/2008_11_21_ccb_standards_fact_sheet_print.pdf (downloaded)
Climate Neutral Network (CN Net)	About - http://www.climateneutral.unep.org/cnn_ contentlist.aspx?m=53 Map of participants - http://www.unep.org/cn_net_map/
Climate Savers	Tokyo Declaration (Feb 2008) - http://www.worldwildlife. org/climate/WWFBinaryitem7178.pdf Company Factsheets (April 2008) - http://www.worldwildlife. org/climate/WWFBinaryitem8752.pdf (downloaded)
ClimateWise	About - http://www.climatewise.org.uk/about-climatewise/ Principles: http://climatevise.squarespace.com/storage/ climatewise-docs/ClimateWise%20Factsheet.pdf (down-loaded) FAQ -http://www.climatewise.org.uk/faqs/
Clinton Climate Initiative	Main page - http://www.clintonfoundation.org/what-we-do/clinton-climate-initiative/ Accomplishments - http://www.clintonfoundation.org/ what-we-do/clinton-climate-initiative/what-we-ve-accomplished
Community Carbon Reduction Project (CRed)	What is Cred? - http://www.cred-uk.org/centralcontent. aspx?intCID=1 2005-06 Report (most recent available) - http://www.cred-uk.org/pdf/CRedFinalReport06.pdf
Conference of New England Governors and Eastern Canadian Premiers Climate Change Action Plan (CNEG/ECP)	Climate Action Plan (2001) - http://www.negc.org/ documents/NEG-ECP%20CCAP.pdf
Connected Urban Development Programme	Overview - http://www.cisco.com/web/about/ac79/docs/ wp/cud/CUD_Overarching_0214final.pdf
Cool Counties Climate Stabilization Initiative	Main site - http://www.kingcounty.gov/exec/coolcounties. aspx Cool Counties Stabilization Declaration - http://www. metrokc.gov/exec/news/2007/0716dec.aspx

continued

Table A.1 (continued)

Experiment	Web site(s)
Covenant of Mayors	About - http://www.eumayors.eu/about_the_covenant/index_en.htm
	Final Covenant of Mayors - http://www.eumayors.eu/mm/staging/library/CoM_text_layouted/Texte_Convention_EN.pdf (downloaded)
	Draft Covenant (Jan 11 2008) - http://www.managenergy.net/download/com/COVENANTDRAFT_11Jan2008.pdf
e8 Network of Expertisefor the Global Environment	Mission Statement - http://www.e8.org/index.jsp?numPage=42
	The e8 Member Companies - http://www.e8.org/index.jsp?numPage=46
Edenbee	About - http://edenbee.com/about
EUROCITIES Declaration on Climate Change	The Declaration - http://www.eurocities.eu/uploads/load.php?file=090215_Eurocities_Declaration_4_Langages-MROD.pdf
Evangelical Climate Initiative	Evangelical Call to Action - http://christiansandclimate.org/learn/call-to-action/
	Act Today - http://christiansandclimate.org/act/
Global GHG Register (GHGR)	Public Reports - http://www.ghgr.org/public/
ICLEI Cities for Climate Protection (CCP) Campaign	CCP: An International Campaign - http://www.iclei.org/index.php?id=1651
	How CCP Works - http://www.iclei.org/index.php?id=810
	Participants - http://www.iclei.org/index.php?id=809
Institutional Investors Group on Climate Change	IIGCC Constitution - http://www.iigcc.org/constitution.aspx
	IIGCC Pamphlet - http://www.iigcc.org/docs/PDF/Public/AboutIIGCC.pdf
International Climate Action Partnership	ICAP Declaration (Oct. 20007) - http://www.icapcarbonaction.com/declaration.htm
	Members - http://www.icapcarbonaction.com/members.htm
	FAQ: http://www.icapcarbonaction.com/faqs.htm
Investor Network on Climate Risk	INCR Membership Info - http://www.incr.com/NETCOMMUNITY/Page.aspx?pid=293&srcid=293
	About INCR - http://www.incr.com/NETCOMMUNITY/Page.aspx?pid=261&srcid=230

Experiment	*Web site(s)*
Investors Group on Climate Change (Australia and New Zealand)	Who are we? - http://www.igcc.org.au/who_are_we Membership - http://www.igcc.org.au/membership
Klimatkommunerna (a.k.a. Swedish Network of Municipalities on Climate Change)	http://www.klimatkommunerna.infomacms.com/
Major Economies Forum on Energy and Climate	White House Fact Sheet May 2007 - http://georgewbush-whitehouse.archives.gov/news/releases/2007/09/20070927.html Joint Statement July 2008 - http://georgewbush-whitehouse.archives.gov/news/releases/2008/07/20080709-5.html Remarks by Secretary Clinton at April 2009 meeting - http://www.state.gov/secretary/rm/2009a/04/122240.htm
Memoranda of Understanding on Climate Change initiated by the State of California	2008 Governor's Global Climate Summit - http://www.climatechange.ca.gov/events/2008_summit/ http://gov.ca.gov/index.php?/press-release/11112/ 2007 MOU - http://gov.ca.gov/pdf/press/070506_climate_change_document.pdf 2008 MOU on deforestation w/ Brazil and Indonesia - http://www.gov.ca.gov/press-release/11101/ 2008 MOU with Mexican border states - http://www.gov.ca.gov/press-release/10419/ 2007 MOU with BC
Methane to Markets	About - http://www.methanetomarkets.org/about/index.htm Partners and Members - http://www.methanetomarkets.org/partners/index.htm Terms of Reference 2004 (incl. signatures until May 2009) - http://www.methanetomarkets.org/join/docs/termsofreference_signed.pdf
Midwestern Greenhouse Gas Reduction Accord	GHG Accord (Nov 15 2007) - http://www.midwesterngovernors.org/Publications/Greenhouse%20gas%20accord_Layout%201.pdf
National Association of Counties (NACo) County Climate Protection Program	FAQ -http://www.naco.org/Template.cfm?Section=New_Technical_Assistance&template=/ContentManagement/ContentDisplay.cfm&ContentID=26515

continued

Table A.1   (continued)

Experiment	Web site(s)
Network of Regional Governments for Sustainable Development (nrg4SD)	Standing Rules of the nrg4SD Statutes (2007) - http://www.nrg4sd.net/pags/AP/AP_Descargas/descarga.asp?id=RE2EE38EC4-C5A8-457F-B00B-AD5DA3A3671B | D2DE27CA-6FCD-442B-BC2C-079D6EFEDF45 Statutes for Legal Registration of the Network (2005) - http://www.nrg4sd.net/pags/AP/AP_Descargas/descarga.asp?id=RE2EE38EC4-C5A8-457F-B00B-AD5DA3A3671B | F4253994-E171-4428-8778-5165163EBC6D
North-South Climate Change Network	CAP site - http://www.cleanairpartnership.org/north_south_climate_change
Ontario-Quebec Provincial Cap-and-Trade Initiative	Press release - http://www.premier-ministre.gouv.qc.ca/salle-de-presse/communiques/2008/juin/2008-06-02-en.shtml MOU June 2008 -http://www.premier-ministre.gouv.qc.ca/salle-de-presse/communiques/2008/juin/2008-06-02-en.shtml
Regional Greenhouse Gas Initiative (RGGI)	RGGI Model Rule - http://www.rggi.org/about/history/model_rule MOUs and amendments - http://www.rggi.org/about/history/mou Fact Sheet – http://www.rggi.org/docs/RGGI_Executive%20Summary_4.22.09.pdf
Renewable Energy and Energy Efficiency Partnership (REEEP)	About REEEP - http://www.reeep.org/48/about-reeep.htm Mission Statement - http://www.reeep.org/511/mission.htm
Southwest Climate Change Initiative	Agreement - http://www.nmclimatechange.us/ewebeditpro/items/O117F8087.pdf (downloaded)
The Climate Group	About - http://www.theclimategroup.org/about The Climate Group Principles - http://theclimategroup.org/assets/The%20%C2%BAClimate%20Group%20Principles.pdf
The Climate Registry (TCR)	FAQ - http://www.theclimateregistry.org/about/faqs.php Principles and Goals - http://www.theclimateregistry.org/principlesgoals.html

Experiment	Web site(s)
Transition Towns	Transition Towns Primer - http://transitiontowns.org/ TransitionNetwork/Primer (incl. picture of Yoda on pg 27!)
Union of the Baltic Cities Resolution on Climate Change	Resolution - http://www.ubc.net/plik,959.html IX General Conference - http://www.ubc.net/ documentation,56,387.html
US-China Memorandum of Understanding to En-hance Cooperation on Climate Change, Energy and the Environment	Press release - http://www.state.gov/r/pa/prs/ps/2009/ july/126592.htm PDF of MOU - http://www.state.gov/documents/ organization/126802.pdf
US Mayors Climate Pro-tection Agreement	The Agreement -http://www.usmayors.org/climate protection/documents/mcpAgreement.pdf
UK Bilateral Climate Change Agreements with US States	UK Michigan Partnership - http://ukinusa.fco.gov.uk/ resources/en/pdf/5581739/uk-michigan-partnership UK Wisconsin Partnership - http://ukinusa.fco.gov.uk/ resources/en/pdf/5581739/Partnership-between-uk-wisconsin UK Florida Partnership - http://ukinusa.fco.gov.uk/ resources/en/pdf/5581739/fl-uk-partnership UK California Partnership - http://ukinusa.fco.gov.uk/ resources/en/pdf/5581739/climate-accord-uk-ca UK Virginia Partnership - http://ukinusa.fco.gov.uk/ resources/en/news/2009/February/uk-virginia-cc-statement#
West Coast Governors' Global Warming Initiative	Main page - http://www.ef.org/westcoastclimate/
Western Climate Initiative (WCI)	Original agreement Feb 2007 http://www.westernclimate initiative.org/ewebeditpro/items/O104F12775. Statement of Regional Goals (August 22, 2007) - http:// www.westernclimateinitiative.org/ewebeditpro/items/ O104F13006.pdf
World Business Council for Sustainable Devel-opment (WBCSD)	About WBCSD - http://www.wbcsd.org/templates/ TemplateWBCSD5/layout.asp?type=p&MenuId=NjA&do Open=1&ClickMenu=LeftMenu FAQ - http://www.wbcsd.org/templates/TemplateWBCSD1/ layout.asp?type=p&MenuId=Mjk0&doOpen=1&ClickMe nu=LeftMenu

Table A.2 **Coding of Climate Governance Experiments***

Experiment	Year Initiated	Market Orientation	Response Orientation	Core Function: Networking	Core Function: Planning	Core Function: Action	Core Function: Oversight
2degrees	2008	999	3	1	0	0	0
Alliance for Resilient Cities	2007	999	2	1	0	0	0
American Carbon Registry	1997	2	1	0	1	0	0
American College & University Presidents Climate Commitment (ACUPCC)	2007	1	1	1	1	1	0
Asia Pacific Partnership on Clean Development and Climate (APP)	2005	3	1	1	0	0	0
Australia's Bilateral Climate Change Partnerships	2002	3	1	1	1	1	0
Business Council on Climate Change (BC3)	2007	1	1	1	1	0	0
C40 Cities Climate Leadership Group	2005	1	3	1	1	1	0
California Climate Action Registry (CCAR)	2001	2	1	0	1	0	1
Carbon Disclosure Project	2000	1	1	0	1	0	0

Experiment	Year Initiated	Market Orientation	Response Orientation	Core Function: Networking	Core Function: Planning	Core Function: Action	Core Function: Oversight
Carbon Finance Capacity Building Programme	2009	1	1	1	0	1	0
Carbon Rationing Action Groups	2005	1	1	1	1	1	1
Carbon Sequestration Leadership Forum	2003	1	1	1	1	0	0
CarbonFix	2007	1	1	0	1	0	0
Chicago Climate Exchange (CCX)	2003	1	1	1	1	1	1
Climate Alliance of European Cities with Indigeouns Rainforest Peoples	1990	999	1	1	1	1	0
Climate, Community, and Biodiversity Alliance (CCBA)	2003	999	1	0	1	0	1
Climate Neutral Network (CN Net)	2008	3	1	1	1	0	0
Climate Savers	1999	1	1	1	1	1	1
ClimateWise	2007	1	3	1	1	1	0
Clinton Climate Initiative	2006	1	1	1	0	1	0

continued

Table A.2 (continued)

Experiment	Year Initiated	Market Orientation	Response Orientation	Core Function: Networking	Core Function: Planning	Core Function: Action	Core Function: Oversight
Community Carbon Reduction Project (CRed)	2003	999	1	1	1	0	0
Conference of New England Governors and Eastern Canadian Premiers Climate Change Action Plan (CNEG/ECP)	2001	3	3	1	1	1	0
Connected Urban Development Programme	2006	1	1	1	0	1	0
Cool Counties Climate Stabilization Initiative	2007	3	3	1	1	1	0
Covenant of Mayors (Europe)	2008	999	1	1	1	1	1
e8 Network of Expertise for the Global Environment	1992	1	1	1	1	1	0
Edenbee	2008	999	1	1	0	0	0
EUROCITIES Declaration on Climate Change	2008	3	3	1	1	1	0
Evangelical Climate Initiative	2006	999	1	1	0	0	0
Global GHG Register (GHGR)	2004	2	1	0	1	0	1

Experiment	Year Initiated	Market Orientation	Response Orientation	Core Function: Networking	Core Function: Planning	Core Function: Action	Core Function: Oversight
ICLEI Cities for Climate Protection (CCP) Campaign	1993	999	3	1	1	0	0
Institutional Investors Group on Climate Change	2001	1	3	1	1	0	0
International Carbon Action Partnership (ICAP)	2007	1	1	1	0	0	0
Investor Network on Climate Risk	2003	1	1	1	1	0	0
Investors Group on Climate Change	2005	1	3	1	1	0	0
Klimatkommunerna	2003	2	1	1	1	0	0
Major Economies Forum on Energy and Climate	2007	3	3	1	0	1	0
Memoranda of Understanding on Climate Change initiated by the State of California	2006	3	3	1	0	1	0
Methane to Markets	2004	1	1	1	1	0	0
Midwestern Greenhouse Gas Reduction Accord	2007	3	1	1	1	1	1

continued

Table A.2 (continued)

Experiment	Year Initiated	Market Orientation	Response Orientation	Core Function: Networking	Core Function: Planning	Core Function: Action	Core Function: Oversight
National Association of Counties (NACo) County Climate Protection Program	2007	3	1	1	1	0	0
Network of Regional Governments for Sustainable Development (nrg4SD)	2002	999	3	1	0	0	0
North-South Climate Change Network	2009	999	2	1	0	0	0
Ontario-Quebec Provincial Cap-and-Trade Initiative	2008	3	1	1	1	1	1
Regional Greenhouse Gas Initiative (RGGI)	2005	3	1	1	1	1	1
Renewable Energy and Energy Efficiency Partnership (REEEP)	2004	1	1	1	0	0	0
Southwest Climate Change Initiative	2006	3	1	1	1	0	0
The Climate Group	2004	3	1	1	0	0	0
The Climate Registry (TCR)	2007	2	1	0	1	0	1
Transition Towns	2005	999	3	1	1	0	0

Experiment	Year Initiated	Market Orientation	Response Orientation	Core Function: Networking	Core Function: Planning	Core Function: Action	Core Function: Oversight
Union of the Baltic Cities Resolution on Climate Change	2007	3	3	1	1	0	0
US-China Memorandum of Understanding to Enhance Cooperation on Climate Change, Energy and the Environment	2009	3	3	1	0	1	0
US Mayors Climate Protection Agreement	2005	3	1	1	1	1	0
UK Bilateral Climate Change Agreements with US States	2006	3	3	1	0	1	0
West Coast Governors' Global Warming Initiative	2004	3	1	1	1	1	0
Western Climate Initiative (WCI)	2007	3	1	1	1	1	1
World Business Council for Sustainable Development (WBCSD)	1990	999	1	1	0	0	0

* For Market Orientation (1=Market, 2=Regulatory, 3=Mixed, 999=indeterminate); For Response Orientation (1=Mitigation, 2=Adaptation, 3=Mixed; For Core Functions (1=activities in core function undertaken, 0=activities in core function not undertaken).

Interviews Undertaken for Project
(Position Titles at Time of Interview)

1. Rohit Aggarwala, Director, Office of Sustainability and Long-Term Planning, New York City Mayor's Office (November 4, 2009).
2. David Antonioli, CEO, Voluntary Carbon Standard (January 20, 2010).
3. Julianne Baroody, Coordinator, Cross-Divisional Climate Change Initiative, Rainforest Alliance (January 19, 2010).
4. Nicholas Bianco, Senior Associate, World Resources Institute (October 29, 2009).
5. Jennifer Biringer, Manager, North American Forest and Trade Network, World Wildlife Fund (September 12, 2005).
6. Derik Broekhof, Vice President of Policy, Climate Action Reserve (February 5, 2010).
7. Robyn Camp, Vice President of Programs, The Climate Registry (January 22, 2010).
8. Alex Carr, Regional Director Canada, The Climate Registry (January 22, 2010).
9. Jonathon Dickinson, Senior Policy Advisor, CC and Greenhouse Gas Inventory, New York City Mayor's Office (November 4, 2009).
10. Jill Duggin, Senior Fellow, World Resources Institute (October 28, 2009).
11. Michele Dyer, Vice President, Strategy and Operations, Second Nature, American College and University Presidents Climate Commitment (November 5, 2009)
12. Georges Dyer, Visiting Senior Fellow, Second Nature, American College and University Presidents Climate Commitment (November 5, 2009).
13. Joanna Durbin, Director, Climate Change and Biodiversity Alliance (October 28, 2009).
14. Jan Franke, Policy Officer – Knowledge Society, EUROCITIES (April 21, 2010, follow-up June 2010)
15. Mary Grady, Director of Membership and Registry Services, American Carbon Registry (October 16, 2009).

16. Jane Gray, State and Regions Programme Director, The Climate Group (November 12, 2009).
17. Callum Grieve, Director of External Affairs North America, The Climate Group (October 9, 2009).
18. Allison Hannon, Midwest Regional Manager, The Climate Group (November 9, 2009).
19. Patrick Hogan, Regional Policy Coordinator, Pew Center on Global Climate Change (October 28, 2009).
20. Peter Holzaepfel, Corporate Engagement Manager, The Climate Group (November 4, 2009).
21. Ulricke Jannssen, Executive Director, Climate Alliance (December 11, 2009).
22. Toby Janson-Smith, Former Director, Climate, Community and Biodiversity Alliance (November 18, 2009).
23. Phil Jessup, Cities and Technology Program Director, The Climate Group (November 13, 2009).
24. Franz Litz, Senior Fellow, World Resources Institute (October 29, 2009).
25. Franz Litz, Senior Fellow, World Resources Institute (January 19, 2010).
26. Mary MacDonald, Director of Environment Office, City of Toronto (October 20, 2009).
27. Megan Meaney, Director Canada Office, International Council for Local Environmental Initiatives (September 15, 2009).
28. Shane Mitchell, Program Manager, Connected Urban Development Program (January 26, 2010).
29. Joseph Pallant, CEO, Carbon Project Solutions (January 26, 2010).
30. Jennifer Penney, Director of Research, Clean Air Partnership (September 15, 2009).
31. Annie Petsonk, International Counsel, Environmental Defense (September 2, 2005).
32. Dasha Rettew, Manager, Cities and Technology, The Climate Group (April 19, 2010, follow-up August 2010).
33. Olivia Ross, Public Relations Director, Clinton Climate Initiative (October 13, 2009; e-mailed questions).
34. Steve Schille, Vice-Chair, Board of Directors, CCAR, Founding Steering Committee, TCR, TCR/CCAR (February 9, 2010).
35. Lisa Scott, Project Coordinator Canada, International Council for Local Environmental Initiatives (July 14, 2008).
36. John Shea, Executive Director, New England Governors' Association (November 6, 2009).
37. Sarah Skikne, Corporate Engagement Manager, The Climate Group (November 16, 2009).

38. Patrick Spencer, Director Internet Business Solutions Group, Cisco (January 26, 2010).

39. Martin Stadelmann, Project Leader, Carbon Offset Project, MyClimate (December 2007).

40. Nicola Villa, Managing Director, Connected Urban Development Program (January 26, 2010, follow up June 2010).

41. Sarah Wade, Partner, AJW (formerly economist at Environmental Defense Fund) (October 27, 2009).

42. Michael Walsh, Executive Vice President, Chicago Climate Exchange (November 24, 2009, follow-up August 2010).

43. Molly Webb, Head of Strategic Engagement, The Climate Group (April 13, 2010).

Notes

Chapter 1

1. The epigraph to this chapter is from Japan 2006. An adapted version of this opening section appeared in Bernstein et al., 2010.
2. UNFCCC 2009.
3. The convening of activities and conferences in parallel to the UN negotiations is a familiar dynamic in climate change and beyond. See Rosenau 2005.
4. Side events are presentations by organizations officially recognized as observers by the United Nations Framework Convention on Climate Change Secretariat, but they are not official components of the negotiations.
5. Interview with Jane Gray; see also www.theclimategroup.org/our-news/news/2009/12/15/climate-leaders-summit/.
6. Grubb 2004; Aall et al. 2007; Auer 2000; Jagers and Stripple 2003; Bulkeley 2005; Hoffmann 2005.
7. In no way should the disappointment with the multilateral process reflect poorly on the professionals who have dedicated their careers and lives to achieving a successful global treaty. Their efforts have kept climate change at the top of the international agenda and spurred on much of the activity discussed in this book.
8. Many of the commitments were contingent. For instance the European Union pledged a 30% reduction from 1990 levels if comparable pledges were made, 20% in the absence of complementary commitments. The United States pledged a 17% reduction from 2005 levels, contingent on action in the U.S. Congress. See UNFCCC 2009.
9. Hoffmann 2007.
10. Dimitrov 2010; Lynas 2009; CBC 2009.
11. Depledge 2006; Hoffmann 2007; Barrett 2003; McKibben and Wilcoxen 2002; Victor 2004; 2006; Schelling 2002.
12. Intergovernmental Panel on Climate Change 2007.
13. See list of CRAGs worldwide at www.carbonrationing.org.uk/groups?country=global (as of May 2010).
14. Quotation from testimonial page on CRAG Web site: www.carbonrationing.org.uk/glasgow/threads/who-are-we.
15. See, e.g., Paterson 2001; Jagers and Stripple 2003; Falkner 2003.
16. For an introduction to Climate Wise see www.climatewise.org.uk/storage/climatewise-docs/ClimateWise%20Factsheet.pdf.
17. For information on the number and characteristics of corporations that report to CDP see https://www.cdproject.net/en-US/WhatWeDo/Pages/overview.aspx.
18. For an introduction to the Investor Network on Climate Risk, see http://www.incr.com//Page.aspx?pid=261&srcid=230

19. See, e.g., Kolk, Levy, and Pinske 2008; Kolk and Pinske 2008.
20. www.iclei.org/index.php?id=809.
21. www.iclei.org/index.php?id=810.
22. www.iclei.org/index.php?id=811.
23. Betsill and Bulkeley 2004, p. 151.
24. Betsill and Bulkeley 2004; Kern and Bulkeley 2009; Lindseth 2004.
25. Bulkeley 2005, p. 893.
26. E.g., Rabe 2004; 2008; Byrne et al. 2007; Moser 2007; Mazmanian et al. 2008.
27. Selin and VanDeveer 2005; Soleille 2006; Betsill and Hoffmann 2008.
28. www.asiapacificpartnership.org/About.htm; www.state.gov/g/oes/climate/mem/. Note that the Major Economies Forum has gone through a number of name changes. It was previously known as the Major Emitters Forum and the Major Economies Process on Energy Security and Climate Change.
29. Lawrence 2007; Kellow 2006; Mcgee and Taplin 2006.
30. Lawrence 2007. See also www.asiapacificpartnership.org/APPTaskforces.htm.
31. International Energy Agency 2009.
32. See Rittel and Webber 1973 for the original discussion of wicked problems and Bernstein et al. 2007 and Prins and Rayner 2007 for discussions of super wicked problems.
33. Bernstein et al. 2007. On path dependence see Pierson 2000; Arthur 1994.
34. Rittel and Webber 1973, p. 163.
35. Hoffmann 2005
36. Hoffmann 2005; Dowdeswell and Kinley 1994; Barrett 1992; Bodansky 1994. Choosing a megamultilateral mode of governing climate change was a political judgment rather than a scientifically obvious way to approach climate change. The scientific understanding of the problem—that it is caused and potentially has effects everywhere and that both the causes and effects are unequally distributed—provides very little means, in and of itself, for narrowing the number of potential resolutions to the problem. Instead, received wisdom, mostly from dealing with the ozone depletion issue in the 1980s (Hoffmann 2005; Downie 1995; Sebenius 1994; Betsill and Pielke 1998) and the dominant position of nation-states in the hierarchy of authority relative to other actors determined the nature of the world's response to climate change.
37. Auer 2000; Raustalia 1997; Newell 2000; Betsill 2002; Betsill and Correll 2008.
38. Bulkeley and Betsill 2003.
39. For a more in-depth discussion of the UNFCCC and Kyoto Protocol negotiations, see Hoffmann 2005.
40. Bodansky 1994; Grubb 1993; Intergovernmental Panel on Climate Change 1990; Jager and Ferguson 1991.
41. Grubb 1993; Rowlands 1995; Harris 2000.
42. Chameides 2009. Of course many individual states will miss their targets (e.g., Canada), and the significance of the accomplishment of the 5% reduction is muted because of the lack of U.S. ratification, the absence of restrictions of major developing states, and the fact that most "reductions" are artifacts of the baseline year (1990) and the collapse of the communist bloc economies that followed in the early 1990s.
43. Sell 1996; Victor 2004; 2006; Barrett 1992; 2003; Wilcoxen and McKibben 2002.
44. European Union action was significantly cheaper because the EU could take advantage of internal variation in carbon dioxide emissions and some fortuitously timed fuel switching from coal to natural gas in the United Kingdom. However, the EU states have also committed to solving the problem of climate change much more stridently than any other set of actors. See Hoffmann 2005.
45. E.g., Grubb 2004.
46. Depledge 2006, pp. 1, 3. See also Bang et al. 2007; Christoff 2006; Victor 2004; 2006; Schelling 2002.
47. For information on the Bali roadmap see the UNFCCC website: http://unfccc.int/meetings/cop_13/items/4049.php.
48. For information on Poznan, see the UNFCCC Web site: http://unfccc.int/meetings/cop_14/items/4481.php

49. See, e.g., Benedick 2007; Christoff 2006; Esty 2006; 2007; Grub 2004; Bodansky 2004; Bodansky and Diringer 2007; Aldy and Stavins 2007; Victor 2006; 2007; Aldy, Barrett, and Stavins 2003; Barrett and Stavins 2003; Buchner and Carraro 2005; Haas 2008; Stewart and Wiener 2008. There are, of course, exceptions to this general rule. See, e.g., Prins and Rayner 2007; Betsill and Bulkeley 2004; Rabe 2004; 2008; Biermann and Pattberg 2010; Newell and Paterson 2010.
50. Prins and Rayner 2007.
51. Ba and Hoffmann 2005; Hewson and Sinclair 1999; Rosenau and Czempiel 1992; Murphy 2000; Avant, Finnemore, and Sell 2009; van Kersbergen and Waarden 2004.
52. Hoffmann and Ba 2005.
53. Rosenau 1997.
54. Research of this nature cannot be undertaken individually—the process of defining and gathering experiments was carried out jointly with my research assistant Gabriel Eidelman. These three criteria resulted from some trial and error as Gabe and I gathered possible experiments and evaluated them independently on each of the criteria discussed here to arrive at the list of eligible experiments.
55. For example, the European Union's emissions trading system is very innovative, but because it is used in the service of meeting the European Union's Kyoto targets, it is not an experiment.
56. My focus is on examining experiments that are rule-making endeavors in nontraditional political spaces, what Hajer calls policy-making in the absence of a polity (Hajer 2003).
57. The UNFCCC Web site's meetings archive only goes as far back as COP 9, held in 2003.
58. See Table 1 in the Appendix for a listing of Web sites from which information on the experiments was drawn.
59. Gupta et al. 2007, p. 144.
60. E.g., Paterson 2001; Betsill and Bulkeley 2004; Rabe 2004; 2008; Moser 2007; Kolk and Pinske 2007; Kern and Bulkeley 2009.
61. Bulkeley 2005; Adger 2001; Selin and VanDeveer 2005 are notable exceptions.
62. Bernstein 2001.
63. www.davidsuzuki.org/issues/climate-change/science/international-climate-negotiations/history-of-climate-negotiations/.

Chapter 2

1. Allison et al. 2009.
2. See, e.g., Kristof 2005; Revkin 2006; Gray 2010.
3. Broder 2010.
4. Intergovernmental Panel on Climate Change 2007, ch. 9.
5. This book is not the place for a debate over the science of climate change. I personally accept the scientific evidence for anthropogenic climate change, but do acknowledge that the inherent uncertainties in both the climate system and climate science leave room for significant debate, not so much over whether climate change is happening but over how much effort to expend in addressing it.
6. Hulme 2009.
7. The number of emerging technological fixes, geoengineering proposals, and lists of possible individual actions have exploded in recent years along with the emergence of governance experiments chronicled here.
8. The term "disruptive change" was suggested by Daniel Yeo of Water Aid during a presentation at the Transnational Climate Governance Workshop in Durham, United Kingdom, September 2010. This friction effect has also been discussed by Rabe (2007; 2008) with regard to state-federal interactions in U.S. environmental policy and Wapner (1966) in relation to the impact of global civil society on domestic environmental policy.
9. All four did evolve into true governance experiments, independent of the Kyoto Protocol and working directly on making rules for addressing climate change.
10. Global "North" and "South" are used in this book as shorthand for developing and developed countries.

11. This finding could be influenced by bias in the data collection process. Experiments that lack access to the UNFCCC negotiations, that do not have websites, or are not publicized in English were not captured by the data-gathering done for this project. Experiments initiated and implemented entirely within the global South may thus exist, but not captured in the database.

12. This is a similar approach to one that many constructivist scholars take to understand the impact of international norms—it is in the rhetorical action and justification of behavior that norms become observable. See Finnemore 2003.

13. The more detailed discussions in the chapters to follow will examine the internal dynamics and external relationships of select experiments, addressing some of the shortcomings of this broader analysis.

14. The concerns raised about using experiments' websites as the source of data are validity concerns—whether the coding accurately reflects what experiments do. The other concern in a coding exercise like this is reliability—consistent application of a rubric across cases. In pursuit of reliable coding, my graduate research assistant and I independently coded each experiment on the seven dimensions and came to consensus on any disputes. This procedure enhances the reliability of the data, but does not ensure its validity.

15. In delineating the activities that an experiment engages in, the focus was on actions that were judged to be central to the functioning of the initiative (i.e., actions without which the experiment would not be said to exist or actions that participants in an experiment must do to be considered part of the initiative).

16. Refer to the Appendix, table A.2, for the coding of experiments on the core functions.

17. In figure 2.4, the label "Activity" refers to the numbered activities in table 2.2.

18. Hajer 2003.

19. Rosenau 1997.

20. In fact, these multilateral initiatives were created as voluntary initiatives at least in part to avoid the binding nature of international treaty-making.

21. Interview with Jennifer Penney. See also the Clean Air Partnership website: http://www.cleanairpartnership.org/about/cap.

22. Interview with Jane Gray.

23. Rabe 2004; 2008.

24. Interviews with Franz Litz and Nicholas Bianco.

25. Avant, Finnemore, and Sell 2009.

26. Rosenau 2003, p. 275

27. Interviews with Michele Dyer and George Dyer See also the ACUPCC website: http://www.presidentsclimatecommitment.org/about/faqs#17.

28. Interview with Megan Meaney.

29. Interview with Callum Grieve.

30. Quote from the Investor Network on Climate Risk's statement on managing risks: www.incr.com/Page.aspx?pid=219 (accessed October 2009).

31. Bernstein 2001, p. 4. Newell and Paterson 2010 discuss this as the emergence of "climate capitalism."

32. Quote from the Business Council on Climate Change's FAQ page: www.bc3sfbay.org/page/43 (accessed July 2009).

33. Harmes 2006; Clapp and Dauvergne 2005; Levy and Newell 2002; Newell and Paterson 2010.

34. Adger 2001.

35. Paterson 2001; Adger 2001.

36. Interview with Mary MacDonald. Zahram et al. 2008a; 2008b also demonstrate that cities that are potentially more vulnerable to climate change are more likely to join the CCP experiment.

37. UNFCCC 2009.

38. Information from North Carolina State University statistics website: http://faculty.chass.ncsu.edu/garson/PA765/cluster.htm.

39. There are a number of possible procedures for performing the grouping. I chose to use a two-step clustering technique because it works best with the categorical data found in

the experiments database: "When at least one variable is categorical, two-step clustering must be used" (http://faculty.chass.ncsu.edu/garson/PA765/cluster.htm).

40. Cluster instability is a noted problem of this analytic technique, and it essentially means that the membership of derived clusters can change depending on the order of the data (i.e., the order in which the statistical package considers each case in performing the cluster analysis). This is a particular issue with smaller data sets like this one ("small datasets can yield strikingly different cluster patterns for different data sequences") and trials with the cluster analysis demonstrated that the combination of a relatively small data set (n = 58) along with a relatively large number of variables (ten rule activities) did not return stable clusters. http://faculty.chass.ncsu.edu/garson/PA765/cluster.htm.

41. I tested the cluster stability by running the cluster analysis on multiple random orderings of the cases.

42. These categories are externally imposed by this study rather than being conscious choices of the experimenters themselves.

43. The epigraph to this section is from www.theclimategroup.org/our-news/news/2004/4/27/collaboration-and-leadership-will-combat-climate-change-says-pm-tony-blair/.

44. Bulkeley 2005; Castells 1996.

45. Castells 1996.

46. See table A.1 for the websites from which the information on the experiments in box 1 was drawn.

47. Susan Kent, Oxfordshire County Council, quoted on 2degrees "success stories" page: http://www.2degreesnetwork.com/how-we-work/member-success-stories/.

48. Interviews with Jane Gray and Sarah Skikne—discussed more fully in chapter 4.

49. www.theclimategroup.org/about-us/ (accessed January 20, 2010).

50. www.2degreesnetwork.com/how-we-work/some-of-our-networks/ (accessed January 20, 2010).

51. The Turner quotation in the epigraph is from www.cdproject.net/en-US/WhatWeDo/Pages/overview.aspx (accessed February 2010).

52. A greenhouse gas offset has become common jargon in the world of climate change. To offset one's greenhouse gas emissions, one can pay for reductions to take place elsewhere. So for instance to offset the emissions I generate by taking a flight, I can pay an offset producer to invest in wind turbines that would replace coal-fired electricity.

53. See Table A.1 in the appendix for the websites from which the information on the experiments in box 2 was drawn.

54. www.theclimateregistry.org/about/.

55. Kollmuss and Bowell 2006; Bumpus and Liverman 2008; Newell and Paterson 2010; Smith 2007.

56. Interview with Megan Meaney.

57. Betsill and Bulkeley 2004.

58. Interviews with Megan Meaney and Alex Carr.

59. Potoski and Prakash 2009; Prakash and Potoski 2006; Ruggie 2004; Clapp 2005; Gibson 1999; Cashore and Bernstein 2004.

60. Interview with Mary MacDonald.

61. Interview with Steve Schiller.

62. See table A.1 for the websites from which the information on the experiments in box 3 was drawn.

63. See table A.1 for the websites from which the information on the experiments in box 4 was drawn.

64. Betsill and Hoffmann 2008; 2011; Skjærseth and Wettestad 2008; Newell and Paterson 2010.

65. James Rosenau, a pioneering scholar of globalization, is fond of asking his students "Of what is this an instance?" about almost anything they see in the world as a way to push their thinking. I have always found it an unparalleled way to think through and attempt to make sense of the world.

Chapter 3

1. Wendt 1999; Rosenau 2005. The quotation in the epigraph from the Web site of the Redlands Bristol CRAG is from www.carbonrationing.org.uk/user/angelaraffle. The quotations from the 2007 UNFCCC COP at Bali were taken from author observations at side event venues during the meetings.
2. Quoted in Waldrop 1992, p. 318; see also Hoffmann and Riley 2002.
3. The inspiration for this kind of theorizing comes from Mary Durfee and James Rosenau (2000) and their intriguing textbook *Thinking Theory Thoroughly*.
4. Of course, theory also influences what we observe and consider to be going on. See, e.g., Adler 1997; Wendt 1999.
5. On constructivism see, e.g., Wendt 1992; 1999; Adler 1997; Onuf 1998; Ba and Hoffmann 2003; Hoffmann 2005; Finnemore 1996; 2003; Price 1997; Klotz 1995; Yee 1996; Wiener 2004; Sandholtz 2008; Cortell and Davis 2005; Checkel 2001; Acharya 2004. On complexity theory see, e.g., Waldrop 1992; Jervis 1997; Holland 1995; Axelrod 1997; Cederman 1997; 2003; Kauffman 1995; Epstein and Axtell 1996. For work that brings the two together see Lustick, Eidelson, and Midownik 2004; Hoffmann 2003; 2005; Cederman 1997. This is obviously not the only kind of story that could be developed to help make sense of climate governance experimentation. Other explanations would focus more exclusively on the material interests and bargaining or actors, or their embeddedness in the global capitalist economy. Of course, as will be discussed below, these factors are not mutually exclusive to a social approach to climate experimentation.
6. This feedback dynamic has many labels. Constructivists would call it mutual constitution or refer to Anthony Giddens's notion of the duality of structure (Giddens 1984). Complexity theorists discuss the feedback as coevolution and emergence.
7. Ruggie 1998.
8. Complexity theorists discuss this as actors' internal rule models—subjective understandings of the world around them. See Holland 1995.
9. Denemark and Hoffmann 2008.
10. Ibid, p. 196.
11. Yee 1996.
12. The process, whether it is called the norm life cycle in constructivism or phase transitions in complexity theory, proceeds in a similar fashion over time; Finnemore and Sikkink 1998; Hoffmann 2005; Bak and Chen 1991; Smith and Stacy 1997.
13. For an interesting argument about lock in see Arthur 1994.
14. Sheingate 2003, p. 191.
15. Constructivism would tend to focus on endogenous sources of uncertainty like the emergence of norm entrepreneurs or contestation over social norms (Finnemore and Sikkink 1998; Hoffmann 2005; Nadelmann 1990; Sandholtz 2008; Wiener 2004; Van Kersbergen and Verbeek 2007), while complexity theorists would tend to focus on exogenous shocks to the system (Bak and Chen 1991).
16. These are called coordination effects (Pierson 2000). Constructivists also recognize this as the building of a critical mass toward a threshold of norm change (Finnemore and Sikkink 1998).
17. Finnemore and Sikkink (1998) discuss this as a cascade in the norm life cycle and it is understood as self-organized criticality in complexity theory (Bak and Chen 1990; Cederman 2003).
18. In constructivism this is where socialization toward a new governance context being taken for granted takes place. In complexity theory this is discussed as the emergence of an attractor that engenders increasing returns.
19. There is obviously a great deal of nuance when this transition cycle plays out in the world. Systems may stay stable for a very long time. The reason for uncertainty to emerge and the contours of the novelty produced can change from system to system. In addition, whether the new system is substantially different from the old one will depend

on idiosyncratic features of the system in question. Figure 3.2 provides a generic framework for structuring thinking about the specifics of the climate governance context.

20. Auer 2000; Newell 2000; Betsill and Correll 2008.
21. Betsill and Bulkeley 2004.
22. This was true beyond climate change and environmental politics. The 1970s–1990s saw the rise of UN megaconferences and universal participation in treaty-making in a number of issue areas—various human rights conferences and the Law of the Sea are relevant examples.
23. Wight 2006; Guzzini 2000; Wendt 1999.
24. Critical junctures have been explored thoroughly in the historical institutionalist literature on path dependence. See Thelen 2000; Pierson 2000; Mahoney 2000; Couch and Farrell 2004.
25. Multilateralism is an example of a very strong governance context (Denemark and Hoffmann 2008; Ruggie 1993) or what complexity theorists would call a strong attractor. Yet even this governance context can be challenged.
26. The governance context is not solely in the "minds" of actors—they cannot *individually* create the world just by believing (Guzzini 2000, p. 155; see also Manicas 1997, p. 11). Yet that does not mean that actors cannot think about and react to their governance context even as it shapes those thoughts and actions. Giddens notes that actors "routinely and for the most part without fuss—maintain a continuing 'theoretical understanding' of the grounds of their activity" (Giddens 1984, p. 5). Archer calls this an actor's internal conversation and claims that "because they [actors] possess personal identity, as defined by their individual configuration of concerns, they [actors] know what they care about most and what they seek to realize in society" (Archer 2003, p. 130).
27. Hopf 2002; Rosenau 1986.
28. Pouliot 2008, p. 265.
29. Rosenau 1981; Hoffmann 2005.
30. Denemark and Hoffmann 2008.
31. Ibid.
32. E.g., Rosenau 2003; Held and McGrew 2007; Scholte 2000; Mittleman 2000; Sassen 2006.
33. Rosenau 2003, p. 281.
34. Sassen 2006, p. 2.
35. Avant, Finnemore, and Sell 2009.
36. Andonova 2010; Andonova, Betsill, and Bulkeley 2009; Rosenau 1990; 1997; Hewson and Sinclair 1999; Ba and Hoffmann 2005; Hall and Biersteker 2002; Haufler 2003; Wapner 1996; Ruggie 2004; Bulkeley 2005; Falkner 2003; Hooghe and Marks 2003.
37. Litfin 2000, quoted in Jagers and Stripple 2003, p. 389.
38. Sinclair 2003; see also Haufler 2003.
39. Avant 2005.
40. Dreher, Molders, and Nunnenkamp 2010; Werker and Ahmed 2008.
41. Andonova, Betsill, and Bulkeley 2009; Cashore and Bernstein 2004; Betsill and Bulkeley 2004; Rabe 2007; 2008; Selin and VanDeveer 2007; Prakash and Potoski 2006.
42. This downloading dynamic has been noted both internationally and in European and North American domestic spheres. See Marwell 2004; Mahon 2005; Weir, Wolman, and Swanstrom 2005.
43. Of course it is possible that the governance context is never really stable, per se, but metastable. It might not be a strict breakpoint, but stability is perceived when centripetal dynamics dominate centrifugal ones—even when both exist simultaneously. Rosenau dubs such simultaneity "fragmegration"—the joining of integrating and fragmenting forces. Rosenau 1997; 2003.
44. Interviews with Sarah Wade and Mary Grady.
45. Betsill and Hoffmann 2008; Skjærseth and Wettestad 2008.
46. Lindseth 2004.

47. See the text of the agreement at www.usmayors.org/climateprotection/documents/mcpAgreement.pdf.
48. Rosenau 1997; 2003, pp. 303 and 296.
49. In addition to constructivism and complexity theory, this translation of uncertainty into experimentation has been extensively explored within organization theory, but the insights from this approach are readily applicable to global governance. Organizational theory models tend to begin with the very dissatisfaction with the system that is evident in the climate governance context. Smith and Stacey (1997) posit that this uncertainty is most likely to occur when an organizational context becomes "brittle," and Gemill and Smith (1985) discuss the subsequent innovation process as "unfreezing" (p. 755).
50. Chapters 4–6 do explore these motivations for specific cases. However, the motivation to experiment in specific cases is not really the target of explanation. The fact of experimentation and its implications is the more important concern for this book. Future research will engage with the motivation to experiment.
51. This list is drawn from Bulkeley, Hoffmann, VanDeveer, and Henson 2010.
52. Personal observations at Toronto Mayors' Event and UNFCCC COP at Bali, December 2007.
53. Bulkeley 2005; Lindseth 2004.
54. How big a change this entails is the key question. It is a new governance system merely because experiments are now functioning. Whether we are likely to see a wholesale replacement of multilateral approaches, the rise to dominance of a particular type of experimental governance, or some hybrid of the two remains to be seen.
55. Comfort 1994, p. 396.
56. Smith and Stacey 1997, p. 85.
57. In constructivist studies of norm dynamics, norm entrepreneurs are considered to draw on existing normative ideas when suggesting new norms. See Finnemore and Sikkink 1998. Complexity theorists discuss the adaptive process as one involving new arrangements of preexisting "building blocks." See Holland 1995 and Sheingate 2003, p. 193.
58. Betsill and Hoffmann 2011
59. Smith and Stacey 1997, p. 80.
60. This point is developed further in chapter 6.
61. Interview with Franz Litz. See also www.westernclimateinitiative.org/news-and-updates/91-regional-greenhouse gas-initiatives-hold-second-meeting.
62. Ibid., p. 397; see also Kauffman 1995.
63. Comfort 1994, p. 397.
64. Victor 2007; Esty 2007.
65. Tiebout 1956.
66. Kollman, Miller, Page 1997.
67. To be sure. the failure at Copenhagen was not positive for the morale of almost anyone concerned with climate change, but the only forward momentum seems to be among experimental initiatives.

Chapter 4

1. The source of the quotation in the epigraph is an author observation from the 2009 UNFCCC COP at Copenhagen, during an ICLEI side event, recorded in notes.
2. Tandon 2010.
3. Quoted in Plummer 2010.
4. This is not a full-scale test of the theoretical story developed in chapter 3, precisely because the individual example cases discussed were also part of the data set that was broadly used to construct the theoretical framework to begin with. Instead, they demonstrate the plausibility of the framework at both the macro and micro scales.
5. The general outline of the origins, functions, and interactions of The Climate Group presented here draws on interviews with a number of The Climate Group staff, including

Jane Gray, Callum Grieve, Allison Hannon, Peter Holzaepfel, Phil Jessup, and Sarah Skikne. It also draws on press releases, annual reports, and papers produced by The Climate Group. See www.theclimategroup.org.

6. Interview with Jane Gray and Northrup 2003.
7. Interview with Jane Gray.
8. Northrup 2003, p. 11.
9. Blair's Speech at The Climate Group Launch (www.number10.gov.uk/Page5716).
10. Ibid.
11. See the press release from The Climate Group launch, found at http://www.the climategroup.org/our-news/events/2004/4/1/the-climategoup-uk-launch/.
12. Blair's Speech at The Climate Group Launch; emphasis added (www.number10.gov.uk/Page5716).
13. Announcement by the Rockefeller Brothers Fund that can be found at www.rbf.org/close_ups/close_ups_show.htm?doc_id=703018, emphasis added.
14. Interview with Callum Grieve.
15. Quote from a report on The Climate Group by Article 13, a corporate social responsibility consulting group: www.article13.com/A13_ContentList.asp?strAction=GetPublication&PNID=938.
16. Interview with Jane Gray.
17. For the text of the law see www.arb.ca.gov/cc/ab32/ab32.htm (accessed February 2010).
18. Climate Group 2006.
19. Interview with Jane Gray.
20. Greenwashing may still occur, and thus a key part of the research agenda for moving forward is to examine why specific corporations join specific experimental initiatives.
21. Interviews with Sarah Skikne and Molly Webb.
22. Interview with Peter Holzaepfel.
23. Interview with Phil Jessup.
24. www.theclimategroup.org/programs/lightsavers/ (accessed January 2010).
25. Interview with Phil Jessup.
26. Interview with Phil Jessup. See also www.theclimategroup.org/programs/lightsavers/ (accessed January 2010). At some point The Climate Group may move beyond the bounds of the Networking model and evolve into the voluntary actor or accountable actor model.
27. Interview with Dasha Rettew—a report publicly announcing these findings is due out in December 2010.
28. www.theclimategroup.org/programs/lightsavers/ (accessed January 2010).
29. Northrup 2003, p. 11
30. Interview with Callum Grieve.
31. www.theclimategroup.org/our-news/events/2005/10/1/world-cities-leadership-climate-change-summit/ (accessed January 2010).
32. Interview with Phil Jessup.
33. Interview with Toby Janson-Smith, press release.
34. Interview with Mary Grady.
35. Interview with Alison Hannon.
36. Interview with Jane Gray.
37. Interview with Jane Gray. See also The Climate Group's report on the UNFCCC COP at Copenhagen: www.theclimategroup.org/our-news/news/2009/12/15/climate-leaders-summit/.
38. Interview with Allison Hannon.
39. Interview with Sarah Skikne, report on breaking deadlock.
40. www.theclimategroup.org/programs/states-and-regions/ (accessed December 2009).
41. Interview with Jane Gray.
42. Interview with Dasha Rettew.
43. Interview with Phil Jessup.
44. www.climateregistry.org/about.html (accessed January 2010).

45. www.climateregistry.org/about/press/press-releases.html (accessed January 2010).
46. Interview with Steve Schiller, and this report on the 2005 UNFCCC COP at Montreal from National Public Radio report, www.npr.org/templates/story/story.php?storyId=5045389 (accessed March 2010).
47. Interview with Steve Schiller.
48. Interview with Steve Schiller.
40. National Public Radio Report, www.npr.org/templates/story/story.php?storyId=5045389 (accessed March 2010).
50. Climate Registry2007.
51. Interview with Steve Schiller.
52. Interview with Alex Carr.
53. The Climate Registry issued a series of press releases from 2007 to 2009 laying out the individual protocols. Direct emissions are those generated through on-site activities. Indirect emissions are those generated when electricity is purchased. www.theclimateregistry.org/news-and-events/press-releases/.
54. Interview with Robyn Camp. See also the Climate Registry 2009.
55. Interview with Robyn Camp.
56. US EPA 2009a; 2009b.
57. US EPA 2009a, p. 11.
58. US EPA 2009a; 2009b.
59. US EPA 2009b, p. 104.
60. US EPA 2009a, p. 11.
61. The Climate Registry is included as a recommended inventory for WCI reporting. See (www.westernclimateinitiative.org/wci-committees/reporting-committee.
62. Midwestern Governors Association 2009.
63. www.theclimateregistry.org/government-services/mandatory-reporting/ (accessed March 2010).
64. Interview with Robyn Camp.
65. The CEO of the Carbon Disclosure Project and the Executive Director of the Climate Registry are both on the board (The Climate Group is represented as well). For more on the Climate Disclosure Standards Board see their 2010 Workplan at: www.cdsb-global.org/uploads/tim. . ./Work%20Plan,%20April%202010.pdf.
66. Interview with Robyn Camp.
67. Parts of this section are drawn from Bulkeley, Hoffmann, VanDeveer, and Henson 2010.
68. Interviews with Phil Jessup, Mary MacDonald.
69. Interviews with Phil Jessup, Mary MacDonald.
70. Interview with Mary MacDonald.
71. Quote taken from C40 report on found at: www.c40cities.org/docs/summit2005/communique.pdf.
72. Quote taken from C40 report found at: www.c40cities.org/news/news-20090519–2jsp (accessed May 21, 2009).
73. Bulkeley and Betsill 2003.
74. Schroeder and Bulkeley 2009.
75. Quote taken from C40 report found at: www.c40cities.org/news/news-20090519–2jsp, (accessed May 21, 2009).
76. Text of the communiqué can be found at: www.kk.dk/Nyheder/2009/December/~/media/B5A397DC695C409983462723E31C995E.ashx.
77. Interview with Mary MacDonald.
78. These initiatives are outlined on the C40 web site at: www.c40cities.org/solutions/.
79. These initiatives are outlined on the C40 web site at: www.c40cities.org/solutions/.
80. These initiatives are outlined on the C40 web site at: www.c40cities.org/initiatives.
81. Interviews with Mary MacDonald and Rohit Aggarwala.
82. Interview with Rohit Aggarwala.
83. Irvine 2009.
84. Osborne 1990; Rabe 2004.
85. Quote from C40 description of its activities at: www.c40cities.org/climatechang.jsp.

86. Quote from C40 description of its activities at: www.c40cities.org/about/goals.jsp.
87. Alison Hannon recalls that the Climate Group sponsored and worked on the C40 summits through 2007 while the Clinton Climate Initiative was developing—personal communication, August 2010.
88. www.climatesummitformayors.dk/.
89. Nordhaus 2008; Stern 2006; Betsill and Hoffmann 2008.
90. The auction debate is a very contentious one—see Betsill and Hoffmann 2010.
91. Raufner and Feldman 1987; Voß 2007.
92. Bernstein, Betsill, Hoffmann, and Paterson 2010.
93. Engels 2006; Andresen and Agarwala 2002; Yamin 1998.
94. Skjærseth and Wettestad 2008; Betsill and Hoffmann 2011.
95. Interview with Michael Walsh. See also www.chicagoclimatex.com/content.jsf?id=1.
96. Combating global warming: study on a global system of tradeable carbon emission entitlements. Geneva: United Nations Conference on Trade and Development.
97. Interview with Michael Walsh.
98. www.unctad.org/Templates/webflyer.asp?docid=3312&intItemID=2280&lang=1.
99. Interview with Michael Walsh.
100. Interview with Michael Walsh.
101. www.chicagoclimatex.com/content.jsf?id=1.
102. Betsill and Hoffmann 2011.
103. Interview with Michael Walsh.
104. Information for this paragraph was distilled from the CCX documents www.chicagoclimatex.com/content.jsf?id=74 and www.chicagoclimatex.com/content.jsf?id=72.
105. The CCX's verification protocols were developed in part through reliance on methods developed by the World Business Council on Sustainable Development, another climate governance experiment, and the World Resources Institute.
106. Interview with Michael Walsh.
107. Sandor, Walsh, and Marques 2002, p. 1893.
108. Interview with Michael Walsh.
109. Interview with Michael Walsh.
110. The plan to cease trading on CCX at the end of Phase II was announced in a press release on October 21, 2010: http://www.chicagoclimatex.com/news/pdf/CCX_Fact_Sheet_20101021.pdf.
111. Mitchell 2008; Young, King, and Schroeder 2008.

Chapter 5

1. The quotation in the epigraph to this chapter is from Betsill and Bulkeley 2007, p. 453.
2. E.g., Bulkeley and Betsill 2003; Betsill and Bulkeley 2004; 2006; Toly 2008; Kousky and Schneider 2003; Stewart 2008; Bulkeley and Kern 2006; 2009; Roman 2010; Osofsky and Levit 2008; Collier 1997.
3. See the C40 Claim at www.c40cities.org/climatechange.jsp.
4. The C40 Seoul Summit Declaration that includes this claim can be found at: www.c40cities.org/news/news-20090522.jsp.
5. The World Mayors and Local Governments Climate Protection Agreement from 2007 (www.globalclimateagreement.org/index.php?id=10395).
6. UN Habitat Program (www.unhabitat.org/content.asp?typeid=19&catid=550&cid=5156).
7. Byrne et al. 2007; Rabe 2008; Kousky and Schneider 2003; Stewart 2008; Betsill and Bulkeley 2004; Koehn 2008.
8. Interview with Rohit Aggarwala.
9. Interview with Molly Webb.
10. U.S. Conference of Mayors 2008.
11. Stewart 2008.
12. Ibid., p. 698.
13. Byrne et al. 2007.

14. Betsill and Bulkeley 2004; Bulkeley and Kern 2006; 2009; Roman 2010; Betsill and Bulkeley 2006; Osofsky and Levit 2008.
15. Betsill and Bulkeley 2004; Roman 2010; Kern and Bulkeley 2009.
16. For the text of the communiqué from which this quote taken see: www.kk.dk/Nyheder/2009/December/~/media/B5A397DC695C409983462723E31C995E.ashx.
17. Kern and Bulkeley 2009, p. 311.
18. Ibid., p. 313; see also Bulkeley and Kern 2006.
19. Betsill and Bulkeley 2006, p. 143; see also Toly 2008.
20. Kern and Bulkeley 2009, p. 323.
21. See, e.g., Allman, Fleming, and Wallace 2004.
22. U.S. Conference of Mayors Survey Report 2008. This is up from 28% that had assessed emissions in the 2007 survey (134 respondents). U.S. Conference of Mayors 2007, www.usmayors.org/climateprotection/climatesurvey07.pdf.www.usmayors.org/climateprotection/documents/2008%20CP%20Survey.pdf.
23. Interview with Megan Meaney; see also ICLEI Canada 2009.
24. ICLEI Canada 2009, pp. 1, 3–4, www.iclei.org/fileadmin/user_upload/documents/Global/Progams/CCP/CCP_Reports/ICLEI_FCM_Canada_2009.pdf.
25. ICLEI Canada 2009, p. 24.
26. Toly 2008.
27. Both Bernstein and Cashore (2000) and Selin and VanDeveer (2006) explore the pathways through which subnational efforts can have a transformative effect on national and global markets.
28. Abate 2006, Wiener 2007, Kousky and Schneider 2003; Aall and Lindseth 2007; Betsill 2001; Deangelo and Harvey 1998; Parker and Rowlands 2007; Osofsky and Levit 2008.
29. Interview with Dasha Rettew.
30. Young 2007, p. 397.
31. Remember that because of the restrictive operational definition from chapter 1, this may be the minimum of municipal activity—it does not capture the individual actions of thousands of cities.
32. This information is from a report on best practices from the U.S. Conference of Mayors: www.usmayors.org/pressreleases/uploads/ClimateBestPractices061209.pdf.
33. Ibid.
34. See the World Business Council for Sustainable Development Manifesto for Energy Efficiency in Buildings: Implementation Guide: www.wbcsd.org/DocRoot/Imq7CBXsnPx-2lqgjNCCu/EEBManifesto.pdf.
35. This information is from a report on best practices from the U.S. Conference of Mayors, www.usmayors.org/pressreleases/uploads/ClimateBestPractices061209.pdf.
36. Interview with Mary MacDonald.
37. Interview with Dasha Rettew.
38. Interview with Molly Webb.
39. Interviews with Dasha Rettew, Molly Webb, Nicola Villa.
40. Interview with Dasha Rettew.
41. Interview with Dasha Rettew. See also (e.g.) a news report from an LED trade magazine: http://www.ledsmagazine.com/news/6/12/14, a press release from the Department of Transportation in New York City: http://www.nyc.gov/html/dot/html/pr2009/pr09_043.shtml, and a press release from the Hong Kong University of Science and Technology: http://www.ust.hk/eng/news/press_20100428-762.html.
42. Interview with Nicola Villa.
43. Ibid.
44. Interview with Jan Franke.
45. Climate Group 2005.
46. Climate Wise is an experiment initiated by the World Wildlife Foundation that seeks corporate pledges to reduced greenhouse gas emissions.
47. For more on the impact and dynamics of overlapping networks see Kern and Bulkeley 2009
48. Interview with Molly Webb.

49. Betsill and Bulkeley 2004, p. 489.
50. www.theclimategroup.org/our-news/news/2009/9/28/cisco-and-the-climate-group-to-develop-new-connected-urban-development-alliance/.
51. Cisco 2006. This quote is from a Cisco press release: http://newsroom.cisco.com/dlls/2006/ts_092106.html.
52. Not only was Cisco in from the beginning on climate change, but they also pledged $10 million over 4 years for the Clinton Foundation's antipoverty efforts in Sub-Saharan Africa. Cisco 2007. See: www.cisco.com/web/about/ac227/csr2008/cisco-and-society/human-needs/sub-saharan-africa-commitment.html.
53. Interview with Nicola Villa.
54. Ibid.
55. For information on Citynet see: http://fibresystems.org/cws/article/magazine/37080.
56. Interview with Nicola Villa. See also Wagener 2008 (www.cisco.com/web/about/ac79/docs/wp/ctd/connected_infra.pdf); Frye 2008 (www.cisco.com/web/about/ac79/docs/wp/ctd/connected_energy.pdf); Kim et al. 2008 (www.cisco.com/web/about/ac79/docs/wp/ctd/connected_mobility.pdf).
57. Interview with Nicola Villa.
58. Ibid.
59. Ibid.
60. For more information on the pilot projects see: www.connectedurbandevelopment.org/pdf/CUD_Program_Overview_2010.pdf
61. For more information on the pilot projects see: www.clintonfoundation.org/what-we-do/clinton-climate-initiative/our-approach/cities/.
62. EUROCITIES 2009 press release: (www.eurocities.eu/main.php?content=content/press/releases.php).
63. See EUROCITIES informational material: www.eurocities.eu/include/lib/sql_document_card.php?id=9609.
64. Interview with Jan Franke.
65. Ibid.
66. Climate Group 2008, p. 9.
67. Ibid, p. 9.
68. Interview with Molly Webb.
69. Ibid.
70. Ibid.
71. Interview with Nicola Villa.
72. Ibid.
73. Ibid.
74. Interview with Molly Webb.
75. Interview with Molly Webb. Molly stressed that Cisco already does a good job of reporting, but that there would be value added in having third party verification of the projects' benefits.
76. Interview with Nicola Villa. For more information see: http://www.cisco.com/web/strategy/smart_connected_communities.html.
77. Interviews with Phil Jessup and Dasha Rettew.
78. Interview with Dasha Rettew.
79. Interview with Molly Webb; Betsill and Bulkeley 2004.
80. Interview with Nicola Villa.
81. Ibid.
82. Interview with Dasha Rettew.
83. Interview with Molly Webb.
84. This raises the key question, to be addressed in chapter 7, of the role of other levels of government. Dasha Rettew noted to me that in places like China and Europe where city initiatives are more integrated into national strategies, the funding issue is less severe and progress can come more quickly.
85. Dietz et al. 2009.
86. Interview with Nicola Villa.

87. See Betsill and Bulkeley 2004; Bulkeley and Betsill 2003 for more on the potential tension between economic and climate change goals.
88. Interview with Nicola Villa.

Chapter 6

1. The quotation in the epigraph is from a 2007 interview with Richard Sandor published at: www.carbon-bus.com/pdfs/Interview_with_Richard_Sandor.pdf (summer 2007).
2. Newell and Paterson 2010; Smith 2007; Stern 2006; Stavins 2008; Nordhaus 2008; Prins and Rayner 2007.
3. Hoffmann 2005; Andresen and Agarwala 1998.
4. Bernstein 2001.
5. Newell and Paterson 2010; Raufner and Feldman 1987; Voß 2007; Engels 2006; Paterson 2010.
6. Bernstein 2001; Paterson 2010.
7. It should be noted that credit and allowance markets are not the only market mechanisms championed in the global response to climate change. The other major possibility is the implementation of various forms of carbon taxes. Economists tend to favor carbon taxes, but they have gained little political traction. For an overview of the debate between cap and trade and carbon taxes see Nordhaus 2008; Stern 2006.
8. There were also significant sustainable development goals embedded in the CDM—this initiative was designed to draw investment from developed countries into projects in the global South that would have simultaneous carbon reduction and development functions.
9. Bernstein et al. 2010; see also Betsill and Hoffmann 2011.
10. This section draws significantly on research undertaken with Michele Betsill. See Betsill and Hoffmann 2011. Venues were identified through a review of reports from Point Carbon and the International Emissions Trading Association, organizations dedicated to providing news, analysis, and consulting services related to the global carbon market, as well as news sources such as the *New York Times* and Environment & Energy Publishing. Graduate and undergraduate research assistants collected data for each venue from a variety of sources such as official Web sites (where they exist), media reports, and secondary sources.
11. See Betsill and Hoffmann 2011.
12. See Betsill and Hoffmann 2011 for full analysis of this variation.
13. Nicholas Bianco reports, "RGGI does not limit projects to the RGGI region. Instead, they can be outside the RGGI states so long as the state the project is located in has a cap-and-trade program and/or has signed an MOU with the RGGI states" (Interview with Nicholas Bianco). For the RGGI policy on offsets see: http://rggi.org/docs/mou_amendment_8_31_06.pdf
14. In 2008 the CDM market was worth $28 billion. See Capoor and Ambrosi 2009.
15. Measurement, verification, and reporting of emissions and emissions reductions are also a key challenge in the cap and trade systems discussed above.
16. Interview with Mary Grady.
17. www.americancarbonregistry.org/membership/Duke%20Energy%20Invests%20in%20GreenTrees.pdf.
18. Interview with Mary Grady. See also Environmental Capital Press Release, www.americancarbonregistry.org/membership/Enviro%20Capital%20sells%20offsets%20to%20Google.pdf.
19. Capoor and Ambrosi 2009, p.1. See also Bernstein et al. 2010.
20. Interview with Mary Grady.
21. Point Carbon is a news and analysis organization focused on carbon markets.
22. Author observation at Point Carbon conference, Carbon Market Insights America November 2009, New York.
23. In most instances, only alternative energy/energy efficiency projects from outside the United States can be used to develop offset credits because of regulations on efficiency

and alternative energy that would make such projects fail the additionality criteria. Interview with Mary Grady.

24. Kollmuss and Bowell 2006.
25. Personal communication with Voluntary Carbon Standard representative at Point Carbon Conference, Carbon Market Insights America November 2009, New York.
26. Interview with Joseph Pallant.
27. Author Observation at Point Carbon Conference, Carbon Market Insights America, November 2009, New York.
28. For more on this see the Chicago Climate Futures web site: www.ccfe.com/.
29. These are experiments directly engaged in the design and functioning of credit and allowance markets. Other experiments have also been labeled market oriented and are indeed engaged in market mechanisms, but they do not directly participate in the credit or allowance markets.
30. This is a point developed further in Betsill and Hoffmann 2008; 2011.
31. New Jersey 1998.
32. Hoffman 2006; Scott 1998.
33. Skjaerseth and Wettestad 2008, p. 68.
34. Interview with David Antonioli.
35. Interview with Franz Litz.
36. Governor quotes from Western Climate Initiative press release: www.westernclimateinitiative.org/component/remository/func-startdown/11/.
37. See, e.g., Lövbrand et al. 2009; Olsen and Painuly 2002; Michaelowa and Jotzo 2005.
38. Interview with Toby Janson-Smith.
39. Interview with David Antonioli.
40. Interviews with Toby Janson-Smith and David Antonioli. See also Kollmuss and Bowell 2006; Bumpus and Liverman 2008; Paterson 2010; Bachram 2004; Smith 2007.
41. A point developed further in Betsill and Hoffmann 2008.
42. Regional Greenhouse Gas Initiative 2005.
43. The Midwestern Accord, WCI, and RGGI met in July and November 2009. Interviews with Patrick Hogan, Nicholas Bianco, and Franz Litz.
44. Author observation at panel at Point Carbon conference, Carbon Market Insights America, November 2009, New York.
45. Bernstein et al. 2010, p. 166.
46. Author observation at the UNFCCC COP in Copenhagen, December 2009—side events on carbon markets sponsored by the International Emissions Trading Association.
47. Ecoplan/Natsource 2006; Jaffe and Stavins 2007; Nordhaus and Danish 2003; Soleille 2006; Stavins 2008.
48. A special issue of *Climate Policy* raised this concern as well. See Flachsland et al. 2009; Grubb 2009; Sterk and Kruger 2009, Tuerk et al. 2009.
49. For a comprehensive analysis of North American climate policies, see Selin and VanDeveer 2009.
50. On auction results see www.rggi.org/co2-auctions/results. On use of the funds generated see www.rggi.org/news.
51. Arizona and Utah are both members of the WCI, but withdrew from the cap and trade program in January and April 2010, respectively.
52. www.midwesternaccord.org/Website%20banner%20quotes.pdf.
53. www.westernclimateinitiative.org/designing-the-program.
54. www.westernclimateinitiative.org/component/remository/func-startdown/11/.
55. Interview with Franz Litz.
56. Rabe 2004; 2008; interviews with Franz Litz, Nicholas Bianco, Patrick Hogan.
57. U.S. House of Representatives 2009; U.S. Senate 2010.
58. See contributions to Selin and VanDeveer 2009.
59. U.S. Senate 2010.
60. Interviews with Franz Litz, Nicholas Bianco, and Patrick Hogan.
61. Interview with Patrick Hogan.

62. Interview with Franz Litz. See also www.westernclimateinitiative.org/news-and-updates/91-regional-ghg-initiatives-hold-second-meeting.
63. Three Regions Offsets Working Group 2010, p. 6.
64. Interview with Nicholas Bianco.
65. Litz and Bianco 2009.
66. Bianco and Litz 2010.
67. Daley 2010.
68. Bianco and Litz 2010.
69. Chicago Climate Exchange 2009.
70. They are also not alone in setting standards—this has become a major tool of environmental activism, especially in the forestry sector with the Forest Stewardship Council. See Bernstein and Cashore 2000; Bernstein et al. 2007; Cashore 2002; Tollefson, Gale, and Healey 2008.
71. Carbon Fix and the CCX have also engaged in setting carbon offset standards but will not be explicitly addressed here.
72. Information on the Voluntary Carbon Standard double approval process can be found at the Voluntary Carbon Standard's Web site: www.v-c-s.org/docs/VCS-Program-Normative-.Document_Double-Approval-Process_v1.1.pdf).
73. www.climateactionreserve.org/how/protocols/adopted/
74. www.climateactionreserve.org/how/
75. Interview with Mary Grady.
76. Ibid.
77. Ibid.
78. Interviews with Joseph Pallant and Toby Janson-Smith.
79. Interview with David Antonioli.
80. Interview with Toby Janson-Smith.
81. Interview with Julianne Baroody.
82. www.americancarbonregistry.org/aboutus/meet-our-team.
83. www.climateactionreserve.org/resources/faqs/#section1. Because of the top-down nature of the Climate Action Reserve, the reverse is not possible.
84. Entergy Press Release, www.americancarbonregistry.org/membership/ETR-Blue-Source_Project-2010_NR.pdf.
85. Interview with Derik Broekhof.
86. Interview with Mary Grady.
87. Interviews with Derik Broekhof, David Antonioli, Toby Janson-Smith, Mary Grady, Joanna Durbin.
88. Interview with David Antonioli.
89. Interview with Derik Broekhoff.
90. Interview with Mary Grady.
91. Interview with Joanna Durbin.
92. Ibid.
93. Ibid. See also Hamilton et al. 2009.
94. Interview with Joanna Durbin.
95. Betsill and Hoffmann 2011.
96. U.S. House of Representatives 2009.
97. Interview with David Antonioli.
98. Interview with Derik Broekhoff.
99. At the time of this writing, the U.S. Congress is struggling to pass climate legislation, and Australia just announced a delay in the implementation of its ambitious cap and trade program because of difficulties passing the program in its legislature.
100. See Newell and Paterson 2010.
101. For instance, emissions from the European Union decreased due to the financial crisis, significantly weakening the demand for permits in the European Union emissions trading system.

Chapter 7

1. The quotation from Rob Hopkins in the epigraph is fromwww.newscientist.com/article/ mg20527466.000-rob-hopkins-getting-over-oil-one-town-at-a-time.html?DCMP=OTC-rss&nsref=online-news (accessed February 9, 2010).
2. The quotation from David Brooks in the epigraph is from www.nytimes. com/2009/07/24/opinion/24brooks.html?_r=1 (accessed July 24, 2009).
3. The quotation from Gywn Prins and Steve Rayner in the epigraph is from Prins and Rayner 2007, pp. v–vi.
4. Hulme 2009.
5. Bernstein 2001.
6. Prins and Rayner 2007.
7. Prins and Rayner 2007; Hoffmann 2007.
8. See, e.g., Bulkeley 2005; Bernstein et al. 2010; Vogler 2003; Selin and VanDeveer 2009; Andonova, Betsill, and Bulkeley 2009; Gupta et al. 2007.
9. Prins and Rayner 2007; Aldy, Barrett, and Stavins 2003; Barrett and Stavins 2003; Buchner and Carraro 2005; Stewart and Wiener 2003; Aldy and Stavins 2007; Keohane and Victor 2010; Keohane 2010; Keohane and Raustalia 2008; Haas 2008; Vogler 2003.
10. Prins and Rayner 2007.
11. Okereke, Bulkeley, and Schroeder 2009.
12. Bernstein and Cashore 2000.
13. Ibid., p. 68.
14. Selin and VanDeveer 2005.
15. Bernstein 2001.
16. See, e.g., Roberts and Parks 2007; Gardiner 2004; Bernstein 2005; Paterson 2010.
17. Prins and Rayner 2007, pp. 4–5.
18. This is by no means a criticism—people involved in experimentation are rightly concentrating on the functions of their initiative.
19. Interview with Dasha Rettew.
20. Again, whether this is true and whether it holds in the medium to long term in addition to the short term must be the focus of continued scrutiny, academic and otherwise.
21. Interviews with Nicola Villa, Phil Jessup, Molly Web, Jan Franke, Dasha Rettew.
22. For information on this program see: www.lowcarboncities.info/.
23. The mission statement of this experiment with this quote can be found at: www.low carboncities.info/about-us/vision-mission.html.
24. Ibid. As a caveat, it is unclear how far along this project has come.
25. Interviews with Patrick Hogan, Nicholas Bianco, Rohit Aggarwala, and Franz Litz.
26. Interview with Rohit Aggarwala.
27. For the 2010 Carbon Expo program see: www.carbonexpo.com/global/dokumente/ carbon_expo/CE2010_Conf_Program.pdf.
28. Denemark and Hoffmann 2008; Bernstein et al. 2010.
29. Special thanks to Michele Betsill for reminding me of this point.
30. For an argument along these lines see Sanwal 2007.
31. Hoffmann 2005; Benedick 2007; Barrett 2007; Victor 2007; Sebenius 1991. For a counterpoint see Esty 2007.
32. Benedick 2007, p. 38.
33. Interview with Phil Jessup.
34. Rotmans and Loorbach 2009, p. 188.
35. From the movie *Caddyshack*, Warner Brothers 1980.
36. Gemmil and Smith 1985, p. 755.
37. Hoffmann 2007; Wiener 2007.

References

Aall, C., K. Groven, and G. Lindseth. 2007. The Scope of Action for Local Climate Policy: The Case of Norway. *Global Environmental Politics* 7(2): 83–101.

Abate, Randall. 2006. Kyoto or Not, Here We Come: The Promise and Perils of the Piecemeal Approach to Climate Change Regulation in the United States. *Cornell Journal of Law and Public Policy* 15(2): 369–401.

Acharya, Amitav. 2004. How Ideas Spread: Whose Norms Matter? *International Organization* 58(2): 239–75.

Adger, W. N. 2001. Scales of Governance and Environmental Justice for Adaptation and Mitigation of Climate Change. *Journal of International Development* 13(7): 921–31.

Adler, Emanuel. 1997. Seizing the Middle Ground: Constructivism in World Politics. *European Journal of International Relations* 3(3): 319–63.

Aldy, Joseph E., and Robert Stavins, eds. 2007. *Architectures for Agreement: Addressing Global Climate Change in the Post-Kyoto World*. Cambridge: Cambridge University Press.

Aldy, Joseph E., Scott Barrett, and Robert N. Stavins. 2003. Thirteen Plus One: A Comparison of Global Climate Policy Architectures. *Climate Policy* 3(4): 373–97.

Allison, I., N. L. Bindoff, R. A. Bindschadler, P. M. Cox, N. de Noblet, M. H. England, J. E. Francis, N. Gruber, A. M. Haywood, D. J. Karoly, G. Kaser, C. Le Quéré, T. M. Lenton, M. E. Mann, B. I. McNeil, A. J. Pitman, S. Rahmstorf, E. Rignot, H. J. Schellnhuber, S. H. Schneider, S. C. Sherwood, R. C. J. Somerville, K. Steffen, E. J. Steig, M. Visbeck, and A. J. Weaver. 2009. *The Copenhagen Diagnosis, 2009: Updating the World on the Latest Climate Science*. Sydney: University of New South Wales, Climate Change Research Centre.

Allman, L., P. Fleming, and A. Wallace. 2004. The Progress of English and Welsh Local Authorities in Addressing Climate Change. *Local Environment* 9(3): 271–83.

Andonova, Liliana. 2010. Public-Private Partnerships for the Earth. Politics and Patterns of Hybrid Authority in the Multilateral System. *Global Environmental Politics* 10 (2): 25–53.

Andonova, Liliana B., Michele M. Betsill, and Harriet Bulkeley. 2009. Transnational Climate Governance. *Global Environmental Politics* 9(2): 52–73.

Andresen, Steinar, and Shardul Agrawala. 2002. Leaders, Pushers and Laggards in the Making of the Climate Change Regime. *Global Environmental Change* 12(1): 41–51.

Archer, Margaret. 2003. *Structure, Agency and the Internal Conversation*. Cambridge: Cambridge University Press.

Arthur, Brian. 1994. *Increasing Returns and Path Dependence in the Economy*. Ann Arbor: University of Michigan Press.

Auer, M. R. 2000. Who Participates in Global Environmental Governance? Partial Answers from International Relations Theory. *Policy Sciences* 33(2): 155–80.

Avant, Deborah. 2005. *The Market for Force: The Consequences of Privatizing Security*. Cambridge: Cambridge University Press.

Avant, Deborah, Martha Finnemore, and Susan Sell. 2009. *Who Governs the Globe?* Cambridge: Cambridge University Press.

Axelrod, Robert. 1997. *The Complexity of Cooperation: Agent Based Models of Competition and Collaboration.* Princeton, N J.: Princeton University Press.

Ba, Alice, and Matthew J. Hoffmann. 2003. Making and Remaking the World for IR 101: A Resource for Teaching Social Constructivism in Introductory Classes. *International Studies Perspectives* 4(1): 15–33.

———, eds. 2005. Contending Perspectives on Global Governance. London: Routledge Press.

Bachram, Heidi. 2004. Climate Fraud and Carbon Colonialism: The New Trade in Greenhouse Gases. *Capitalism, Nature, Socialism* 15(4): 5–20.

Bak, Per, and Kan Chen. 1991. Self-organized Criticality. *Scientific American* (January): 46–53.

Bang, Guri, Camilla Bretteville Froyn, Jon Hovi, and Fredric C. Menz. 2007. The United States and International Climate Cooperation: International "Pull" versus Domestic "Push." *Energy Policy* 35(2): 1282–91.

Barrett, Scott. 1992. *Convention on Climate Change: Economic Aspects of Negotiations.* Paris: OECD.

———. 2003. *Environment and Statecraft.* Oxford: Oxford University Press.

———. 2007. A Multitrack Climate Treaty System. In *Architectures for Agreement: Addressing Global Climate Change in the Post-Kyoto World,* edited by Joseph E. Aldy and Robert N. Stavins, 237–59. Cambridge: Cambridge University Press.

Barrett, Scott, and Robert Stavins. 2003. Increasing Participation and Compliance in International Climate Change Agreements. *International Environmental Agreements: Politics, Law and Economics* 3(4): 349–76.

Benedick, Richard. 2007. Avoiding Gridlock on Climate Change. *Issues in Science and Technology* 23(2): 37–40.

Bergek, Anna, Staffan Jacobsson, and Bjorn Sanden. 2008. "Legitimation" and "Development of Positive Externalities": Two Key Processes in the Formation of Phase of Technological Innovations Systems. *Technology Analysis and Strategic Management* 20(5): 575–92.

Bernstein, Steven. 2001. *The Compromise of Liberal Environmentalism.* New York: Columbia University Press.

———. 2005. Legitimacy in Global Environmental Governance. *Journal of International Relations and International Law* 1(1–2): 139–66.

Bernstein, Steven, Michele Betsill, Matthew Hoffmann, and Matthew Paterson. 2010. A Tale of Two Copenhagens: Carbon Markets and Climate Governance. *Millennium* 39 (1): 161–73.

Bernstein, Steven, and Benjamin Cashore. 2000. Globalization, Fourth Paths of Internationalization and Domestic Policy Change: The Case of Eco-forestry Policy Change in British Columbia, Canada. *Canadian Journal of Political Science* 33(1): 67–99.

Bernstein, Steven, Benjamin Cashore, Kelly Levin, and Graeme Auld. 2007. Playing It Forward: Path Dependency, Progressive Incrementalism, and the "Super Wicked" Problem of Global Climate Change. Paper presented at the annual meeting of the International Studies Association, Chicago.

Betsill, Michele. 2001. Mitigating Climate Change in US Cities: Opportunities and Obstacles. *Local Environment* 6(4): 393–406.

———. 2002. Environmental NGOs Meet the Sovereign State: The Kyoto Protocol Negotiations on Global Climate Change. *Colorado Journal of International Environmental Law and Policy* 13(1): 49–64.

Betsill, Michele, and Harriet Bulkeley. 2004. Transnational Networks and Global Environmental Governance: the Cities for Climate Protection Program. *International Studies Quarterly* 48(2): 471–93.

———. 2006. Cities and the Multilevel Governance of Global Climate Change. *Global Governance* 12(2): 141–59.

———. 2007. Looking Back and Thinking Ahead: A Decade of Cities and Climate Change Research. *Local Environment* 12(5): 447–56.

Betsill, Michele, and Elizabeth Correll (eds). 2008. *NGO Diplomacy: The Influence of Nongovernmental Organizations in International Environmental Negotiations* Cambridge: MIT Press.

Betsill, Michele, and Matthew J. Hoffmann. 2008. The Evolution of Emissions Trading Systems for Greenhouse Gases. Paper presented at the annual meeting of the International Studies Association, San Francisco, March.

———. 2011. The Contours of Cap and Trade: The Evolution of Emissions Trading Systems for Greenhouse Gases. *Review of Policy Research* 27(1): 81–103.

Betsill, Michele, and Roger Pielke. 1998. Blurring the Boundaries: Domestic and International Ozone Politics and Lessons for Climate Change. *International Environmental Affairs* 10(3): 147–72.

Bianco, Nicholas, and Franz Litz. 2010. Reducing Greenhouse Gas Emissions in the United States Using Existing Federal Authorities and State Action. World Resources Institute Report: http://www.wri.org/publication/reducing-ghg-emissions-using-existing-federal-authorities-and-state-action

Biermann, Frank, Phillip Pattberg, and Fariborz Zelli (eds). 2010. *Global Climate Governance Beyond 2012: Architecture, Agency, and Adaptation.* Cambridge: Cambridge University Press.

Bodansky, Daniel. 1994. Prologue to the Climate Change Convention. In *Negotiating Climate Change: The Inside Story of the Rio Convention*, edited by Irving Mintzer and J. A. Leonard, 45–74. Cambridge: Cambridge University Press.

———. With contributions from Sophie Chou and Christie Jorge-Tresolini. 2004. International Climate Efforts beyond 2012: A Survey of Approaches. Pew Center on Global Climate Change, Washington, DC. www.pewclimate.org/docUploads/2012%20new.pdf.

Bodansky, Daniel, and Eliot Diringer. 2007. Towards an Integrated Multi-track Climate Framework. Pew Center on Global Climate Change, Washington, DC. www.pewclimate.org/multi-track.

Broder, John M. 2010, February 11. Climate-CHANGE Debate Is Heating Up in Deep Freeze. *New York Times.* www.nytimes.com/2010/02/11/science/earth/11climate.html.

Brooks, David. 2009, July 23. Kill the Rhinos! *New York Times.* www.nytimes.com/2009/07/24/opinion/24brooks.html?_r=3.

Buchner, Barbara, and Carlo Carraro. 2005. Economic and Environmental Effectiveness of a Technology-based Climate Protocol. *Climate Policy* 4: 229–48.

Bulkeley, Harriet. 2005. Reconfiguring Environmental Governance: Towards a Politics of Scales and Networks. *Political Geography* 24(8): 875–902.

Bulkeley, Harriet, and Michele M. Betsill. 2003. *Cities and Climate Change: Urban Sustainability and Global Environmental Governance.* London: Routledge.

Bulkeley, Harriet, Matthew Hoffmann, Stacy D. VanDeveer, and Tori Henson. 2010. Transnational Governance Experiments: Evidence from the Climate Change Arena. Manuscript in preparation.

Bulkeley, Harriet, and Kristine Kern. 2006. Local Government and the Governing of Climate Change in Germany and the UK. *Urban Studies* 43(12): 2237–59.

Bumpus, Adam, and Diana Liverman. 2008. Accumulation by Decarbonization and the Governance of Carbon Offsets. *Economic Geography* 84(2): 127–55.

Byrne, John, Kristen Hughes, Wilson Rickerson, and Lado Kurdgelashvili. 2007. American Policy Conflict in the Greenhouse: Divergent Trends in Federal, Regional, State, and Local Green Energy and Climate Change Policy. *Energy Policy* 35(9): 4555–73.

Capoor, Karan, and Philippe Ambrosi. 2009. *State and Trends of the Carbon Market 2009.* Washington, D.C.: World Bank.

Cashore, Benjamin. 2002. Legitimacy and the Privatization of Environmental Governance: How Non-state Market-driven (NSMD) Governance Systems Gain Rule Making Authority. *Governance* 15(4): 503–29.

Cashore, Benjamin, and Steven Bernstein. 2004. Non-state Global Governance: Is Forest Certification a Legitimate Alternative to a Global Forest Convention? In *Hard Choices, Soft Law: Combining Trade, Environment, and Social Cohesion in Global Governance*, edited by John Kirton and Michael Trebilcock, 33–64. Aldershot, England: Ashgate Press.

Castells, Manuel. 1996. *The Rise of the Network Society: The Information Age. Vol. 1 of Economy, Society and Culture.* Cambridge, Mass: Blackwell.

References

CBC. 2009, December 18. Canada Tagged as "Fossil of the Year." CBC News. www.cbc.ca/politics/story/2009/12/18/climate-canada-award.html.

Cederman, Lars-Erik. 1997. *Emergent Actors in World Politics*. Princeton, N.J.: Princeton University Press.

———. 2003. Modeling the Size of Wars: From Billiard Balls to Sandpiles. *American Political Science Review* 97(1): 135–50.

Chameides, Bill. 2009. Did the Kyoto Protocol Miss the Target? *The Huffington Post* (October 12), http://www.huffingtonpost.com/bill-chameides/did-the-kyoto-protocol-mi_b_317855.html

Checkel, Jeffrey. 2001. Why Comply? Social Learning and European Identity Change. *International Organization* 55(3): 553–88.

Chicago Climate Exchange. 2009. Chicago Climate Exchange Offsets Report September-December 1(5). www.chicagoclimatex.com/content.jsf?id=1800.

Christoff, Peter. 2006. Post-Kyoto? Post-Bush? Towards an Effective "Climate Coalition of the Willing." *International Affairs* 82(5): 831–60.

Clapp, Jennifer. 2005. Global Environmental Governance for Corporate Responsibility and Accountability. *Global Environmental Politics* 5(3): 23–34.

Clapp, Jennifer, and Peter Dauvergne. 2005. *Pathways to a Green World: The Political Economy of the Global Environment*. Cambridge, Mass.: MIT Press.

Climate Group, The. 2005, October. Low Carbon Leader: Cities. www.theclimategroup.org/publications/2005/10/1/low-carbon-leader-cities/.

———. 2006. Blair and Schwarzenegger Join International Business Leaders for Energy Roundtable. www.theclimategroup.org/our-news/news/2006/7/31/the-climate-group-convenes-climate-and-energy-roundtable/.

———. 2008. SMART2020: Enabling the Low Carbon Economy in the Information Age. www.theclimategroup.org/publications/2008/6/19/smart2020-enabling-the-low-carbon-economy-in-the-information-age/.

Climate Registry, The. 2009. Members Survey. www.theclimateregistry.org/downloads/2009/05/members_survey_2009.pdf.

Collier, U. 1997. "Local Authorities and Climate Protection in the EU: Putting Subsidiarity into Practice?" *Local Environment* 2(1):39–57.

Comfort, Louise K. 1994. Self-organization in Complex Systems. *Journal of Public Administration Research and Theory* 4(3): 393–410.

Copenhagen Climate Summit for Mayors, The. 2009, December 15. Cities Act: The Copenhagen Climate Communiqué. www.kk.dk/climatesummitformayors.aspx.

Corell, Elisabeth, and Michele M. Betsill. 2001. A Comparative Look at NGO Influence in International Environmental Negotiations: Desertification and Climate Change. *Global Environmental Politics* 1(4): 86–107.

———. 2008. *NGO Diplomacy: The Influence of Non-governmental Organizations in International Environmental Negotiations*. Cambridge, Mass.: MIT Press.

Cortell, Andrew P., and James W. Davis. 2005. When Norms Clash: International Norms, Domestic Practices, and Japan's Internalisation of the GATT/WTO. *Review of International Studies* 31(1): 3–25.

Couch, Colin, and Henry Farrell. 2004. Breaking the Path of Institutional Development? Alternative to the New Determinism. *Rationality and Society* 16(1): 5–43.

Daley, Beth. 2010, April 9. US Climate Bill Weak for N.E., Critics Say. *Boston Globe*. www.boston.com/news/local/massachusetts/articles/2010/04/09/us_climate_bill_weak_for_ne_critics_say.

Deangelo, B. J., and Harvey, L.D.D. 1998. The Jurisdictional Framework for Municipal Action to Reduce Greenhouse Gas Emissions: Case Studies from Canada, the USA and Germany. *Local Environment* 3(2): 111–36.

Denemark, Robert, and Matthew J. Hoffmann. 2008. Not Just Scraps of Paper: The Dynamics of Multilateral Treaty-Making. *Cooperation and Conflict* 43(2): 185–219.

Depledge, Joanna. 2006. The Opposite of Learning: Ossification in the Climate Change Regime. *Global Environmental Politics* 6(1): 1–22.

Dietz, Thomas, Gerald T. Gardner, Jonathan Gilligan, Paul C. Stern, and Michael P. Vandenbergh. 2009. Household Actions Can Provide a Behavioral Wedge to Rapidly Reduce U.S. Carbon Emissions. *Proceedings of the National Academy of Sciences* 106(44): 18452–56.

Dimitrov, Radoslav. 2010. Inside Copenhagen: The State of Climate Governance. *Global Environmental Politics* 10(2): 18–24.

Dowdeswell, Elizabeth, and Richard Kinley. 1994. Constructive Damage to the Status Quo. In *Negotiating Climate Change: The Inside Story of the Rio Convention*, edited by Irving Mintzer and J. A. Leonard, 113–28. Cambridge: Cambridge University Press.

Downie, David Leonard. 1995. Road Map or False Trail? Evaluating the "Precedence" of the Ozone Regime as a Model and Strategy for Global Climate Change. *International Environmental Affairs* 7(4): 321–45.

Dreher, Axel, Florian Molders, and Peter Nunnenkamp. 2010. Aid Delivery through Non-governmental Organisations: Does the Aid Channel Matter for the Targeting of Swedish Aid? *World Economy* 33(2): 147–76.

Durfee, Mary, and James N. Rosenau. 2000. *Thinking Theory Thoroughly: Coherent Approaches to an Incoherent World*. Boulder, Colo.: Westview Press.

Ecoplan/Natsource. 2006. *Linking Domestic Emissions Trading Schemes to the EU ETS*. http://sd-cite.iisd.org/cgi-bin/koha/opac-detail.pl?biblionumber=36157.

Engels, Anita. 2006. Market Creation and Transnational Rule-Making: The Case of CO2 Emissions Trading. In *Transnational Governance: Institutional Dynamics of Regulation*, edited by M. L. Djelic and K. Sahlin-Andersson, 329–74. Cambridge: Cambridge University Press.

Epstein, Joshua, and Robert Axtell. 1996. *Growing Artificial Societies: Social Science from the Bottom Up*. Washington, D.C.: Brookings Institution Press.

Esty, Daniel C. 2006. From Local to Global: The Changing Face of the Environmental Challenge. *SAIS Review* 26(2): 191–97.

———. 2007. Beyond Kyoto: Learning from the Montreal Protocol. In *Architectures for Agreement: Addressing Global Climate Change in the Post-Kyoto World*, edited by Joseph E. Aldy and Robert N. Stavins, 260–69. Cambridge: Cambridge University Press.

Falkner, Robert. 2003. Private Environmental Governance and International Relations: Exploring the Links. *Global Environmental Politics* 3(2): 72–87.

Finnemore, Martha. 1996. *National Interests in International Society*. Ithaca, N.Y.: Cornell University Press.

———. 2003. *The Purpose of Intervention: Changing Beliefs about the Use of Force*. Ithaca, N.Y.: Cornell University Press.

Finnemore, Martha, and Sikkink, Kathryn. 1998. International Norm Dynamics and Political Change. *International Organization* 52(4): 887–918.

Flachsland, Christian, Robert Marschinski, and Ottmar Edenhofer. 2009. To Link or Not to Link: Benefits and Disadvantages of Linking Cap and Trade Systems. *Climate Policy* 9(4): 358–72.

Frye, Wes. 2008. Connected and Sustainable Energy. White paper written for the Connected Urban Development Global Conference, Amsterdam. Available from Cisco's website at: www.cisco.com/web/about/ac79/docs/wp/ctd/connected_energy.pdf.

Gardiner, Stephen M. 2004. Ethics and Global Climate Change. *Ethics* 114: 555–600.

Gemilll, Gary, and Charles Smith. 1985. A Dissipative Structure Model of Organization Transformation. *Human Relations* 38(8): 751–66.

Gibson, Robert B., ed. 1999. *Voluntary Initiatives and the New Politics of Corporate Greening*. Peterborough, Canada: Broadview Press.

Giddens, Anthony. 1984. *The Constitution of Society: Outline of the Theory of Structuration*. Berkeley: University of California Press.

Gray, Louise. 2010. Pakistan floods: Climate change experts say global warming could be the cause. *The Telegraph UK*. 10 August 2010. http://www.telegraph.co.uk/news/worldnews/asia/pakistan/7937269/Pakistan-floods-Climate-change-experts-say-global-warming-could-be-the-cause.html

Grubb, Michael. 1993. *The Earth Summit Agreements: A Guide and Assessment*. London: Earthscan.

———. 2004. Kyoto and the Future of International Climate Change Responses: From Here to Where? *International Review for Environmental Strategies* 5(1): 15–38.

———. 2009. Linking Emissions Trading Schemes. *Climate Policy* 9(4): 339–40.

Gupta, Joyeeta, Kim Van Der Leeuw, and Hans De Moel. 2007. Climate Change: A "Glocal" Problem Requiring "Glocal" Action. *Environmental Sciences* 4(3): 139–48.

Guzzini, Stefano. 2000. A Reconstruction of Constructivism in International Relations. *European Journal of International Relations* 6(2): 147–82.

Haas, Peter M. 2008. Climate Change Governance after Bali. *Global Environmental Politics* 8(3): 1–7.

Hajer, Maarten. 2003. Policy without Polity? Policy Analysis and the Institutional Void. *Policy Sciences* 36(2): 175–95.

Hall, Rodney Bruce, and Thomas Biersteker, eds. 2002. *The Emergence of Private Authority in Global Governance*. Cambridge: Cambridge University Press.

Hamilton, Katherine, Milo Sjardin, Allison Shapiro, and Thomas Marcello. 2009. Fortifying the Foundation: State of the Voluntary Carbon Markets 2009. Report by Ecosystem Marketplace and New Carbon Finance. www.ecosystemmarketplace.com/pages/dynamic/article.page.php?page_id=6773andsection=homeandeod=1.

Harris, Paul, ed. 2000. Climate Change and American Foreign Policy. New York: St. Martin's Press.

Harmes, Adam. 2006. Neoliberalism and Multilevel Governance. *Review of International Political Economy* 13(5): 725–49.

Haufler, Virginia. 2003. Globalization and Industry Self-Regulation. In *Governance in a Global Economy: Political Authority in Transition*, edited by Miles Kahler and David Lake. Princeton: Princeton University Press.

Held, David, and Anthony McGrew, eds. 2007. *Globalization Theory: Approaches and Controversies*. Cambridge: Polity Press.

Hewson, Martin, and Timothy Sinclair, eds. 1999. *Approaches to Global Governance Theory*. Albany: State University of New York Press.

Hoffman, Andrew J. 2006. *Getting Ahead of the Curve: Corporate Strategies That Address Climate Change*. Washington, D.C.: Pew Center on Global Climate Change.

Hoffmann, Matthew J. 2003. Constructing a Complex World: The Frontiers of International Relations Theory and Foreign Policy–Making. *Asian Journal of Political Science* 11(2): 37–57.

———. 2005. *Ozone Depletion and Climate Change: Constructing a Global Response*. Albany: State University of New York Press.

———. 2007. The Global Regime: Current Status of and Quo Vadis for Kyoto. In *A Globally Integrated Climate Policy for Canada*, edited by Steven Bernstein, Jutta Brunnee, David G. Duff, and Andrew J. Greene, 137–57. Toronto: University of Toronto Press.

Hoffmann, Matthew, and Alice Ba. 2005. Introduction: Coherence and Contestation. In *Contending Perspectives on Global Governance: Coherence, Contestation and World Order*, 1–14. London: Routledge.

Hoffmann, Matthew, and John Riley. 2002. The Science of Political Science: Linearity or Complexity in the Design of Social Inquiry. *New Political Science* 24(2): 303–20.

Holland, John. 1995. *Hidden Order*. New York: Addison-Wesley.

———. 1998. *Emergence: From Chaos to Order*. Cambridge: Perseus Books.

Hooghe, Liesbet, and Gary Marks. 2003. Unraveling the Central State, but How? Types of Multilevel Governance. *American Political Science Review* 97(2): 233–43.

Hopf, Ted. 2002. *Social Construction of International Politics: Identities and Foreign Policies, Moscow 1955 and 1999*. Ithaca, N.Y.: Cornell University Press.

Hulme, Mike. 2009. *Why We Disagree about Climate Change: Understanding Controversy, Inaction and Opportunity*. Cambridge: Cambridge University Press.

Intergovernmental Panel on Climate Change. 1990. *Climate Change*. Washington, D.C.: Island Press.

———. 2007. *Contribution of Working Group III to the Fourth Assessment Report of the Intergovernmental Panel on Climate Change*. Edited by B. Metz, O. R. Davidson, P. R. Bosch, R. Dave, and L. A. Meyer. Cambridge: Cambridge University Press. www.ipcc.ch/publications_and_data/ar4/wg3/en/contents.html.

International Council for Local Environmental Initiatives Canada. 2009. Partners for Climate Protection Annual Measures Report 2009. www.iclei.org/fileadmin/user_upload/documents/Global/Progams/CCP/CCP_Reports/ICLEI_FCM_Canada_2009.pdf.

International Energy Agency. 2009. CO2 Emissions from Fuel Combustion: Highlights. 2009 Edition. www.iea.org/co2highlights/co2highlights.pdf.

Irvine, Dean. 2009, May 20. Bill Clinton to Cities: Act on Climate. CNN News. www.cnn.com/2009/TECH/science/05/20/seoul.climate/index.html (accessed May 21, 2009).

Jaffe, Judson, and Robert N. Stavins. 2007. Linking Tradable Permit Systems for Greenhouse Gas Emissions: Opportunities, Implications, and Challenges. Geneva: International Emissions Trading Association. Paper accessed from: http://belfercenter.ksg.harvard.edu/publication/17770/linking_tradable_permit_systems_for_greenhouse_gas_emissions.html.

Jager, Jill, and Howard Ferguson, eds. 1991. *Climate Change: Science, Impacts and Policy: Proceedings of the Second World Climate Conference*. Cambridge: Cambridge University Press.

Jagers, Sverker C., and Johannes Stripple. 2003. Climate Governance beyond the State. *Global Governance* 9(3): 385–99.

Japan. 2006. November. Work by Annex I Parties on the Scientific Basis for Determining Their Further Commitments, Including on Scenarios for Stabilization of Atmospheric Concentrations of GHG and on the Implications of These Scenarios. Japanese Presentation at First In-session Workshop of the Ad Hoc Working Group on Further Commitments for Annex I Parties under the Kyoto Protocol. UNFCCC document available from: http://unfccc.int/files/meetings/cop_12/insession_workshops/application/pdf/061107_4_awg_japan_1.pdf.

Jervis, Robert. 1997. *System Effects*. Princeton, N.J.: Princeton University Press.

Kauffman, Stuart. 1995. *At Home in the Universe*. New York: Oxford University Press.

Kellow, Aynsley. 2006. A New Process for Negotiating Multilateral Environmental Agreements? The Asia-Pacific Climate Partnership beyond Kyoto. *Australian Journal of International Affairs* 60(2): 287–303.

Keohane, Robert O. 2010. The Economy of Esteem and Climate Change. *St Antony's International Review* 5(2): 16–28.

Keohane, Robert O., and Kal Raustalia. 2008. Toward a Post-Kyoto Climate Change Architecture: A Political Analysis. Discussion paper 08–01, Harvard Project on International Climate Agreements. Working paper accessed from: http://belfercenter.ksg.harvard.edu/files/Keohane%20and%20Raustiala%20HPICA1.pdf.

Keohane, Robert O., and David G. Victor. 2010. The Regime Complex for Climate Change. Discussion Paper 10–33, Harvard Project on International Climate Agreements. http://belfercenter.ksg.harvard.edu/publication/19880/regime_complex_for_climate_change.html.

Kern, Kristine, and Harriet Bulkeley. 2009. Cities, Europeanization and Multi-level Governance: Governing Climate Change through Transnational Municipal Networks. *Journal of Common Market Studies* 47(2): 309–32.

Kim, Jayes, Tony Kim, Todd Litman, J.D. Stanley, and Val Stoyanov. 2008. Connected and Sustainable Mobility. White paper written for the Connected Urban Development Global Conference, Amsterdam. Paper accessed from Cisco website: www.cisco.com/web/about/ac79/docs/wp/ctd/connected_mobility.pdf.

Klotz, Audie. 1995. *Norms in International Relations: The Struggles against Apartheid*. Ithaca, N.Y.: Cornell University Press.

Koehn, Peter H. 2008. Underneath Kyoto: Emerging Subnational Government Initiatives and Incipient Issue-bundling Opportunities in China and the United States. *Global Environmental Politics* 8(1): 53–77.

Kolk, Ans, David Levy, and Jonatan Pinkse. 2008. Corporate Responses in an Emerging Climate Regime: The Institutionalization and Commensuration of Carbon Disclosure. *European Accounting Review* 17(4): 719–45.

Kolk, Ans, and Jonatan Pinkse. 2007. Multinationals' Political Activities on Climate Change. *Business and Society* 46(2): 201–28.

———. 2008. Business and Climate Change: Emergent Institutions in Global Governance. *Corporate Governance: International Journal of Business in Society* 8(4): 419–29.

Kollman, Ken, John H. Miller, and Scott E. Page. 1997. Political Institutions and Sorting in a Tiebout Model. *American Economic Review* 87(5): 977–92.

Kollmuss, Anja, and Benjamin Bowell. 2006. Voluntary Offsets for Air-travel Carbon Emissions: Evaluations and Recommendations of Voluntary Offset Companies. Tufts Climate Initiative. www.tufts.edu/tie/tci/pdf/TCI_Carbon_Offsets_Paper_Jan31.pdf.

Kousky, Carolyn, and Stephen H. Schneider. 2003. Global Climate Policy: Will Cities Lead the Way? *Climate Policy* 3(4): 359–72.

Kristof, Nicholas D. 2005, September 11. The Storm Next Time. *New York Times*. www.nytimes.com/2005/09/11/opinion/11kristof.html.

Lawrence, Peter. 2007. The Asia Pacific Partnership on Clean Development and Climate (AP6): a Distraction to the Kyoto Process or a Viable Alternative? *Asia Pacific Journal of Environmental Law* 10(4): 183–209.

Levy, David L., and Peter Newell. 2002. Business Strategy and International Environmental Governance: Toward a Neo-Gramscian Synthesis. *Global Environmental Politics* 3(4): 84–101.

Lindseth, Gard. 2004. The Cities for Climate Protection Campaign (CCPC) and the Framing of Local Climate Policy. *Local Environment* 9(4): 325–36.

Litfin, Karen. 2000. Environment, Wealth, and Authority: Global Climate Change and Emerging Modes of Legitimation. *International Studies Review* 2(2): 119–48.

Litz, Franz, and Nicholas Bianco. 2009. Keeping the Light On in the State Laboratory: Enabling U.S. States to Achieve Greenhouse Gas Emissions Reductions through Retirement of Federal Cap-and-Trade Allowances.World Resources Institute issue brief. www.wri.org/node/11295.

Lövbrand, Eva, Teresia Rindefjäll, and Joakim Nordqvist. 2009. Closing the Legitimacy Gap in Global Environmental Governance? Lessons from the Emerging CDM Market. *Global Environmental Politics* 9(2): 74–100.

Lustick, Ian S., Roy J. Eidelson, and Dan Miodownik. 2004. Secessionism in Multicultural States: Does Sharing Power Prevent or Encourage It? *American Political Science Review* 94(2): 209–30.

Lynas, Mark. 2009, December 22. How Do I Know China Wrecked the Copenhagen Deal? I Was in the Room. *Guardian*. www.guardian.co.uk/environment/2009/dec/22/copenhagen-climate-change-mark-lynas.

Mahon, Rianne. 2005. Rescaling Social Reproduction: Childcare in Toronto/Canada and Stockholm/Sweden. *International Journal of Urban and Regional Research* 29(2): 341–57.

Mahoney, James. 2000. Path Dependence in Historical Sociology. *Theory and Society* 29(4): 507–48.

Manicas, P. 1997. The Concept of Structure. In *Anthony Giddens: Critical Assessments*, vol. 2, edited by Christopher Bryant and David Jary. New York: Routledge.

Marwell, Nicole. 2004. Privatizing the Welfare State: Nonprofit Community-based Organizations and Political Actors. *American Sociological Review* 69(2): 265–91.

Mazmanian, Daniel A., John Jurewitz, and Hal Nelson. 2008. California's Climate Change Policy: The Case of a Subnational State Actor Tackling a Global Challenge. *Journal of Environment and Development* 17(4): 401–23.

Mcgee, Jeffrey, and Ros Taplin. 2006. The Asia-Pacific Partnership on Clean Development and Climate: A Complement or Competitor to the Kyoto Protocol? *Global Change, Peace and Security* 18(3): 173–92.

McKibben, Warwick, and Peter Wilcoxen. 2002. *Climate Change Policy after Kyoto: Blueprint for a Realistic Approach*. Washington, D.C.: Brookings Institution Press.

Michaelowa, Axel. 2006. Principles of Climate Policy after 2012. *Intereconomics* 41(2): 60–63.

Michaelowa, Axel, and Frank Jotzo. 2005. Transaction Costs, Institutional Rigidities and the Size of the Clean Development Mechanism. *Energy Policy* 33(4): 511–23.

Midwestern Governors Association. 2009, June. Midwestern Greenhouse Gas Reduction Accord: Advisory Group Draft Final Recommendations. www.midwesternaccord.org/GHG%20Draft%20Advisory%20Group%20Recommendations.pdf.

Mitchell, Ronald B. 2008. Evaluating the Performance of Environmental Institutions: What to Evaluate and How to Evaluate It? In *Institutions and Environmental Change: Principal Findings, Applications, and Research Frontiers*, edited by Oran R. Young, Leslie A. King, and Heike Schroeder, 79–114. Cambridge, Mass.: MIT Press.

Mittleman, James. 2000. *The Globalization Syndrome: Transformation and Resistance*. Princeton, N.J.: Princeton University Press.

Moser, Susanne. 2007. In the Long Shadows of Inaction: The Quiet Building of a Climate Protection Movement in the United States. *Global Environmental Politics* 7(2): 124–44.

Murphy, Craig. 2000. Global Governance: Poorly Done and Poorly Understood. *International Affairs* 76(4): 789–803.

Nadelman, Ethan. 1990. Global Prohibition Regimes: The Evolution of Norms in International Society. *International Organization* 44(4): 479–526.

Newell, Peter. 2000. *Climate for Change: Non-state Actors and the Global Politics of the Greenhouse.* Cambridge: Cambridge University Press.

Newell, Peter, and Matthew Paterson. 2010. *Climate Capitalism: Global Warming and the Transformation of the Global Economy.* Cambridge: Cambridge University Press.

New Jersey. 1998, June 5. Letter of Intent between Ministry of Housing, Spatial Planning and the Environment, The Netherlands, and Department of Environmental Protection, State of New Jersey. Reprinted in New Jersey, Department of Environmental Protection, Sustainability Greenhouse Action Plan, December 1999, app. D, A11–A12. http://njedl.rutgers.edu/ftp/PDFs/2564.pdf (accessed January 21, 2010).

Nordhaus, Robert R., and Kyle W. Danish. 2003. Designing a Mandatory Greenhouse Gas Reduction Program for the U.S. Washington, D.C.: Pew Center on Global Climate Change. www.pewclimate.org/publications/report/designing-mandatory-greenhouse-gas-reduction-program-us.

Nordhaus, William. 2008. *A Question of Balance.* New Haven: Yale University Press.

Northrup, Michael. 2003. Cutting Greenhouse Gas Emissions is Possible and Even Profitable. Working paper published by Ecologic. Available at http://ecologic.eu/683.

Okereke, Chukwumerije, Harriet Bulkeley, and Heike Schroeder. 2009. Conceptualizing Climate Governance beyond the International Regime. *Global Environmental Politics* 9(1): 58–78.

Olsen, Kim. R., and Jyoti P. Painuly. 2002. The Clean Development Mechanism: A Bane or a Boon for Developing Countries? *International Environmental Agreements: Politics, Law and Economics* 2(3): 237–60.

Onuf, Nicholas. 1998. Constructivism: A User's Manual. In *International Relations in a Constructed World*, edited by Vendulka Kubalkova, Nicholas Onuf, and Paul Kowert, 58–78. Armonk, N.Y.: Sharpe.

Osborne, David. 1990. *Laboratories of Democracy.* Cambridge, Mass.: Harvard Business School Press.

Osofsky, Hari M., and Janet Koven Levit. 2008. The Scale of Networks? Local Climate Change Coalitions. *Chicago Journal of International Law* 8(2): 409–36.

Parker, P., and I. H. Rowlands. 2007. City Partners Maintain Climate Change Action Despite National Cuts: Residential Energy Efficiency Programme Valued at Local Level. *Local Environment* 12(5): 505–17.

Paterson, Matthew. 2001. Risky Business: Insurance Companies in Global Warming Politics. *Global Environmental Politics* 1(4): 18–42.

———. 2010. Legitimation and Accumulation in Climate Change Governance. *New Political Economy* 15(3): 1–23.

Pierson, Paul. 2000. Increasing Returns, Path Dependence, and the Study of Politics. *American Political Science Review* 94(2): 251–68.

Plumer, Bradford. 2010. Is the Real Action on Climate Policy in the States? New Republic Online. www.tnr.com/blog/the-vine/the-real-action-climate-policy-the-states (accessed February 10, 2010).

Potoski, Matthew, and Aseem Prakash, eds. 2009. *Voluntary Programs: A Club Theory Approach.* Cambridge, Mass.: MIT Press.

Pouliot, Vincent. 2008. The Logic of Practicality: A Theory of Practice of Security Communities. *International Organization* 62(2): 257–88.

Prakash, Aseem, and Matthew Potoski. 2006. *The Voluntary Environmentalist? Green Clubs and ISO 14001.* Cambridge: Cambridge University Press.

Price, Richard. 1997. *The Chemical Weapons Taboo.* Ithaca, N.Y.: Cornell University Press.

Prins, Gwyn, and Steve Rayner. 2007. The Wrong Trousers: Radically Rethinking Climate Policy. Joint research paper of James Martin Institute for Science and Civilization and MacKinder Centre for the Study of Long-Wave Events. Oxford: James Martin Institute.

Rabe, Barry G. 2004. *Statehouse and Greenhouse: The Emerging Politics of American Climate Change Policy*. Washington, D.C.: Brookings Institution Press.

———. 2007. Beyond Kyoto: Climate Change Policy in Multilevel Governance Systems. *Governance:An International Journal of Policy, Administration, and Institutions* 20(3): 423–44.

———. 2008. States on Steroids: The Intergovernmental Odyssey of American Climate Policy. *Review of Policy Research* 25(2): 105–28.

Raufner, Roger, and Stephen Feldman. 1987. *Acid Rain and Emissions Trading*. New York: Rowman and Littlefield.

Raustalia, Kal. 1997. States, NGOs and International Environmental Institutions. *International Studies Quarterly* 41(4): 719–40.

Regional Greenhouse Gas Initiative. 2005, 20 December. Memorandum of Understanding. www.rggi.org/docs/mou_12_20_05.pdf.

Revkin, Andrew C. 2006, May 22. "An Inconvenient Truth": Al Gore's Fight against Global Warming. *New York Times*. www.nytimes.com/2006/05/22/movies/22gore.html.

Rittel, Horst W. J., and Melvin M. Webber. 1973. Dilemmas in a General Theory of Planning. *Policy Sciences* 4: 155–69.

Roberts and Parks Roman, Mikael. 2010. Governing from the Middle: The C40 Cities Leadership Group. *Corporate Governance* 10(1): 73–84.

Rosenau, James N. 1981. *The Study of Political Adaptation*. London: Frances Pinter.

———. 1986. Before Cooperation: Hegemons, Regimes, and Habit-driven Actors in World Politics. *International Organization* 40(4): 849–50.

———. 1990. *Turbulence in World Politics: A Theory of Change and Continuity*. Princeton: Princeton University Press.

———. 1997. *Along the Domestic-Foreign Frontier: Exploring Governance in a Turbulent World*. Cambridge: Cambridge University Press.

———. 2003. *Distant Proximities: Dynamics beyond Globalization*. Princeton: Princeton University Press.

———. 2005. Global Governance as Disaggregated Complexity. In *Contending Perspectives on Global Governance*, edited by Alice Ba and Matthew J. Hoffmann. London: Routledge.

Rosenau, James N., and Ernst-Otto Czempiel, eds. 1992. *Governance without Government: Order and Change in World Politics*. Cambridge: Cambridge University Press.

Rotmans, Jan, and Derk Loorbach. 2009. Complexity and Transition Management. *Journal of Industrial Ecology* 13(2): 184–96.

Rowlands, Ian. 1995. *The Politics of Global Atmospheric Change*. New York: Manchester University Press.

Ruggie, John, ed. 1993. *Multilateralism Matters: The Theory and Praxis of an Institutional Form*. New York: Columbia University Press.

———. 1998. What Makes the World Hang Together? Neo-utilitarianism and the Social Constructivist Challenge. *International Organization* 52(4): 1–40.

———. 2004. Reconstituting the Global Public Domain: Issues, Actors and Practices. *European Journal of International Relations* 10(4): 499–531.

Sandholtz, Wayne. 2008. Dynamics of International Norm Change: Rules against Wartime Plunder. *European Journal of International Relations* 14(1): 101–31.

Sandor, Richard, Michael Walsh, and Rafael Marques. 2002. Greenhouse-gas-trading Markets. *Philosophical Transactions of the Royal Society London* 360: 1889–1900.

Sanwal, Mukul. 2007. Evolution of Global Environmental Governance and the United Nations. *Global Environmental Politics* 7(3): 1–12.

Sassen, Saskia. 2006. *Territory, Authority, Rights: From Medieval to Global Assemblages*. Princeton, N.J.: Princeton University Press.

Schelling, Thomas C. 2002. What Makes Greenhouse Sense? Time to Rethink the Kyoto Protocol. *Foreign Affairs* 81(3): 2–9.

Scholte, Jan Aart. 2000. *Globalization: A Critical Introduction*. London: Macmillan.

Schroeder, Heike, and Harriet Bulkeley. 2009. Global Cities and the Governance of Climate Change: What Is the Role of Law in Cities? *Fordham Urban Law Journal* 36: 313–59.

Scott, Alex. 1998. BP Experiments with CO2 Emissions Trading. *Chemical Week* (October 28): 42.

Sebenius, James. 1991. Designing Negotiations toward a New Regime: The Case of Global Warming. *International Security* 15(4): 110–48.

———. 1994. Towards a Winning Climate Coalition. In *Negotiating Climate Change*, edited by Irving Mintzer and J. A. Leonard, 277–320. Cambridge: Cambridge University Press.

Selin, Henrik, and Stacy D. VanDeveer. 2005. Canadian-U.S. Environmental Cooperation: Climate Change Networks and Regional Action. *American Review of Canadian Studies* 35(2): 353–78.

———. 2007. Political Science and Prediction: What's Next for U.S. Climate Change Policy? *Review of Policy Research* 24(1): 1–27.

———, eds. 2009. *Changing Climates in North American Politics: Institutions, Policymaking and Multilevel Governance*. Cambridge, Mass.: MIT Press.

Sell, Susan. 1996. North-South Environmental Bargaining: Ozone, Climate Change, and Biodiversity. *Global Governance* 2(1): 93–116.

Sheingate, Adam. 2003. Political Entrepreneurship, Institutional Change, and American Political Development. *Studies in American Political Development* 17: 185–203.

Sinclair, Timothy. 2003. Global Monitor: Bond Rating Agencies. *New Political Economy* 8(1): 147–61.

Skjærseth, Jon Birger, and Jorgen Wettestad. 2008. *EU Emissions Trading: Initiation, Decision-making and Implementation*. Aldershot, England: Ashgate.

Smith, Kevin. 2007. *The Carbon Neutral Myth: Offset Indulgences for Your Climate Sins*. Amsterdam: Carbon Trade Watch.

Smith, Michaela, and Ralph Stacey. 1997. Governance and Cooperative Networks: An Adaptive Systems Perspective. *Technological Forecasting and Social Change* 54(1): 79–94.

Soleille, Sebastien. 2006. Greenhouse Gas Emissions Trading Schemes: A New Tool for the Environmental Regulator's Kit. *Energy Policy* 34(13): 1473–47.

Stavins, Robert N. 2008. Addressing Climate Change with a Comprehensive U.S. Cap-and-Trade System. Discussion paper 2008–02. Cambridge, Mass.: Belfer Center for Science and International Affairs, John F. Kennedy School of Government. http://belfercenter.ksg.harvard.edu/publication/18203/addressing_global_climate_change_with_a_comprehensive_us_capandtrade_system.html.

Sterk, Wolfgang, and Joseph Kruger. 2009. Establishing a Transatlantic Carbon Market. *Climate Policy* 9(4): 389–401.

Stern, Nicholas. 2006. *The Stern Review Report on the Economics of Climate Change*. Cambridge: Cambridge University Press.

Stewart, Richard B. 2008. States (and Cities) as Actors in Global Climate Regulation: Unitary vs. Plural Architectures. *Arizona Law Review* 50(3): 681–707.

Stewart, Richard B., and Jonathan B. Wiener. 2003. *Reconstructing Climate Policy: Beyond Kyoto*. Jackson, Tenn.: AEI Press.

Tandon, Shaun. 2010, February 9. US Warns China against "Stillborn" Climate Deal. AFP. www.google.com/hostednews/afp/article/ALeqM5jG_Jwc2rEnT30Kwe48QakDv-TofA (accessed February 10, 2010).

Thelen, Kathleen. 2000. Timing and Temporality in the Analysis of Institutional Evolution and Change. *Studies in American Political Development* 14(1): 101–8.

Three Regions Offsets Working Group. 2010. Ensuring Offset Quality: Design and Implementation Criteria for a High-quality Offset Program. www.westernclimateinitiative.org/component/remository/general/Ensuring-Offset-Quality-Design-and-Implementation-Criteria-for-a-High-Quality-Offset-Program/.

Tiebout, Charles. 1956. A Pure Theory of Local Expenditures. *Journal of Political Economy* 64(5): 416–24.

Tietenberg, T. 1992. Relevant Experience with Tradable Permits. In *Combating Global Warming: Study on a Global System of Tradable Carbon Emission Entitlements: United Nations Conference on Trade and Development*, 37–54. New York: United Nations.

Tollefson, Chris, Fred Gale, and David Haley. 2008. *Setting the Standard: Certification, Governance, and the Forest Stewardship Council*. Vancouver:University of British Columbia Press.

Toly, Noah J. 2008. Transnational Municipal Networks in Climate Politics: From Global Governance to Global Politics. *Globalizations* 5(3): 341–56.

Tuerk, Andreas, Michael Mehling, Christian Flachsland, and Wolfgang Sterk. 2009. Linking Carbon Markets: Concepts, Case Studies and Pathways. *Climate Policy* 9(4): 341–57.

United Nations Framework Convention on Climate Change. 2009. Copenhagen Accord. Draft decision-/CP.15 Proposal by the President. http://unfccc.int/resource/docs/2009/cop15/eng/l07.pdf.

U.S. Conference of Mayors. 2007, summer. Survey on Mayoral Leadership on Climate Protection. www.usmayors.org/climateprotection/climatesurvey07.pdf.

——. 2008, June. The Impact of Gas Prices, Economic Conditions, and Resource Constraints on Climate Protection Strategies in U.S. Cities. www.usmayors.org/climateprotection/documents/2008%20CP%20Survey.pdf.

——. 2009, June. Taking Local Action: Mayors and Climate Protection Best Practices. www.usmayors.org/pressreleases/uploads/ClimateBestPractices061209.pdf.

U.S. EPA. 2009a, September. Mandatory Greenhouse Gas Reporting Rule: EPA's Response to Public Comments. Vol. 4. Approach to Verification and Missing Data. www.epa.gov/climatechange/emissions/downloads09/documents/Volume4-Verification-MissingData-FINAL.pdf.

——. 2009b, September. Mandatory Greenhouse Gas Reporting Rule: EPA's Response to Public Comments. Vol. 11. Designated Representative and Data Collection, Reporting, Management and Dissemination. www.epa.gov/climatechange/emissions/downloads09/documents/Volume11-DesignatedRepDataCollection-FINAL.pdf.

U.S. House of Representatives. 2009. H.R. 2454: American Clean Energy and Security Act of 2009. 111th Cong., 1st sess., June 26, 2009. http://energycommerce.house.gov/index.php?option=com_contentandview=articleandid=1633:the-american-clean-energy-and-security-act-of-2009-hr-2454andcatid=169:legislationandItemid=55.

U.S. Senate. 2010. American Power Act—Discussion Draft. http://kerry.senate.gov/americanpoweract/pdf/APABill.pdf.

van Kersbergen, Kees, and Frans van Waarden. 2004. "Governance" as a Bridge between Disciplines: Cross-disciplinary Inspiration Regarding Shifts in Governance and Problems of Governability, Accountability and Legitimacy. *European Journal of Political Research* 43: 143–71.

van Kersbergen, Kees, and Bertjan Verbeek. 2007. The Politics of International Norms: Subsidiarity and the Imperfect Competence Regime of the European Union. *European Journal of International Relations* 13(2): 217–38.

Victor, David. 2004. *Climate Change: Debating America's Policy Options*. New York: Council on Foreign Relations.

——. 2006. Toward Effective International Cooperation on Climate Change: Numbers, Interests and Institutions. *Global Environmental Politics* 6(3): 90–103.

——. 2007. Fragmented Carbon Markets and Reluctant Nations: Implications for the Design of Effective Architectures. In *Architectures for Agreement: Addressing Global Climate Change in the Post-Kyoto World*, edited by Joseph E. Aldy and Robert N. Stavins, 133–60. Cambridge: Cambridge University Press.

Vogler, John. 2003. Taking Institutions Seriously: How Regime Analysis Can Be Relevant to Multilevel Environmental Governance. *Global Environmental Politics* 3(2): 25–39.

Voluntary Carbon Standard. 2010 January 21. VCS Program Normative Document: Double Approval Process. Version 1.1. www.v-c-s.org/docs/VCS-Program-Normative-Document_Double-Approval-Process_v1.1.pdf.

Voß, Jan-Peter. 2007. Innovation Processes in Governance: The Development of "Emissions Trading" as a New Policy Instrument. *Science and Public Policy* 34(5): 329–43.

Wagener, Wolfgang. 2008. Connected and Sustainable ICT Infrastructure. White paper written for the Connected Urban Development Global Conference, Amsterdam. Paper accessed from Cisco web site: www.cisco.com/web/about/ac79/docs/wp/ctd/connected_infra.pdf.

Waldrop, M. Mitchell. 1992. *Complexity: The Emerging Science at the Edge of Order and Chaos*. New York: Simon and Schuster.

Wapner, Paul. 1996. *Environmental Activism and World Civic Politics*. Albany: State University of New York Press.

Weir, Margaret, Harold Wolman, and Todd Swanstrom. 2005. The Calculus of Coalitions: Cities, Suburbs, and the Metropolitan Agenda. *Urban Affairs Review* 40(6): 730–60.

Wendt, Alexander. 1992. Anarchy Is What States Make of It: The Social Construction of Power Politics. *International Organization* 46(2): 391–25.

———. 1999. *Social Theory of International Politics*. Cambridge: Cambridge University Press.

Werker, Eric, and Faisal Z. Ahmed. 2008. What Do Nongovernmental Organizations Do? *Journal of Economic Perspectives* 22(2): 73–92.

Wiener, Antje. 2004. Contested Compliance: Interventions on the Normative Structure of World Politics. *European Journal of International Relations* 10(2): 189–234.

———. 2008. *The Invisible Constitution of Politics: Contested Norms and International Encounters*. Cambridge: Cambridge University Press.

Wiener, Jonathan B. 2007. Think Globally Act Globally: The Limits of Local Climate Politics. *University of Pennsylvania Law Review* 155: 1961–79.

Wight, Colin. 2006. *Agents and Structures in International Relations: Politics as Ontology*. Cambridge: Cambridge University Press.

Yamin, Farhana. 1998. Climate Change Negotiations: An Analysis of the Kyoto Protocol. *International Journal of Environment and Pollution* 10(3–4): 428–53.

Yee, Albert. 1996. The Causal Effects of Ideas on Policies. *International Organization* 50(1): 69–108.

Young, Abby. 2007. Forming Networks, Enabling Leaders, Financing Action: The Cities for Climate Protection Campaign. In *Creating a Climate for Change: Communicating Climate Change and Facilitating Social Change*, edited by S. C. Moser and L. Dilling, 383–98. Cambridge: Cambridge University Press.

Young, Oran R., Leslie A. King, and Heike Schroeder, eds. 2008. *Institutions and Environmental Change: Principal Findings, Applications, and Research Frontiers*. Boston: MIT Press.

Zahran, Sammy, Samuel D. Brody, Arnold Vedlitz, Himanshu Grover, and Caitlyn Miller. 2008b. Vulnerability and Capacity: Explaining Local Commitment to Climate-change Policy. *Environment and Planning C: Government and Policy* 26: 544–62.

Zahran, Sammy, Himanshu Grover, Samuel D. Brody, and Arnold Vedlitz. 2008a. Risk, Stress, and Capacity: Explaining Metropolitan Commitment to Climate Protection. *Urban Affairs Review* 43(4): 447–74.

Index

accountable actors, accountable action, 26, 41–42, 76, 153, 191n26
    activities of, 54–57
    carbon market development and, 124, 134–36
    in practice, 81, 95–98
acid rain, 73, 96, 124–25
action planning, planning, 34–36, 74, 119
    as activity of experiments, 34–35, 41–42, 47–52, 54, 56, 172–76
    C40 and, 92–94, 101
    governance models and, 41–42, 47–52, 54, 56
activity clusters, 26, 77, 79, 86
    carbon market development and, 123–24, 134–49, 153, 159, 197n29
    and potentials of experimentation, 153, 159–60, 163
    and self-organization of experiments, 73, 75
    and technology deployment in cities, 103–4, 107–22, 153, 159
ACUPCC. See American College & University Presidents' Climate Commitment
adaptation, 68
    and activities of experiments, 19, 23, 33
    and challenges of climate change, 10–11
    Copenhagen meetings and, 4–6, 41, 152
    governance models and, 51, 54
    liberal environmentalism and, 36, 38, 40–41
    and self-organization of experiments, 72–73
    See also complex adaptive social construction
Aggarwala, Rohit, 94, 105, 177
Alliance for Resilient Cities, 19, 38, 43, 45, 165, 172
American Carbon Registry, 39, 68, 74, 85, 165, 172, 177
    activities of, 19, 48–49
    in carbon market development, 130–33, 135, 137, 145–47
American College & University Presidents' Climate Commitment (ACUPCC), 38, 52–53, 165, 172, 177
    activities of, 19, 53

Antonioli, David, 138–39, 146, 148–49, 177
Archer, Margaret, 189n26
Asia-Pacific Partnership on Clean Development and Climate (APP), 8, 10, 37, 43, 45, 165, 172
Australia, 22, 71, 169
    carbon market development and, 126–29, 149, 198n99
    governance models and, 49, 53
    megamultilateral development and, 13, 16
Australia's Bilateral Climate Change Partnerships, 19, 165, 172
autos, 47, 104, 107
    fuel efficiency of, 27–28, 54

Bali, UNFCCC COP negotiations in, x, 6, 16, 59, 71
BC3 (Business Council on Climate Change), 19, 40, 48, 165, 172
Bella Conference Center, 3–6, 151–52
Benedict, Richard, 161
Bernstein, Steven, 39, 157
Betsill, Michele, 7, 103, 114
Bianco, Nicholas, 177, 196n13
bilateral partnerships and memoranda of understanding, 53
Blair, Tony, 82–83
Boston, Mass., 111
British Petroleum (BP), 82–83, 88, 97
    carbon market development and, 127, 137, 148
Broekhoff, Derik, 147, 177
Brooks, David, 151
Brown, Lord John, 148
buildings, 19, 94
    carbon market development and, 141, 159
    governance models and, 45, 52–54
    and technology deployment in cities, 109–14
Bulkeley, Harriet, 7, 103, 106, 114
Bush, George W., 15, 77, 152
Business Council on Climate Change (BC3), 19, 40, 48, 165, 172
Byrd-Hegel resolution, 14–15

Cadman, David, 80–81
California, 22, 53
    carbon market development and, 127, 138, 141
    Climate Group and, 82–84
    TCR and, 87–89
California Climate Action Registry (CCAR), 39, 52,
        87–89, 166, 172, 178
    activities of, 19, 48
California Climate Action Reserve. *See* Climate
        Action Reserve
Camp, Robyn, 89, 91, 177
Canada, ix, 6, 18, 28, 32, 55, 92, 178
    and activities of experiments, 22–24
    bottom-up initiatives and, 7–8
    carbon market development and, 8, 127,
        140–41, 143
    and city-level responses to climate change,
        106–7
    governance models and, 49–50, 54, 56
    liberal environmentalism and, 37–39
    megamultilateral development and, 13, 16,
        184n42
    and self-organization of experiments, 74–75
    TCR and, 88–91, 101
cap and trade systems, 4, 8, 28, 76, 80, 123–31
    and activities of experiments, 20, 23–24, 34–35
    and carbon market development, 124–31,
        134–43, 162, 196n13, 196n15, 198n99,
        198n101
    and CCX, 95–101, 123, 135, 144
    and governance models, 55–57
    and Kyoto Protocol, 68, 96–97, 185n55
    and liberal environmentalism, 37–39
    linking of, 140, 143
    multiple venues for, 126–29
    and self-organization of experiments, 73–75
    TCR and, 90–91
capitalism, 40, 188n5
Carbon Disclosure Project, 7, 76, 166, 172
    activities of, 19, 46, 48–49
    liberal environmentalism and, 38–39
carbon emissions, carbon emissions reduction, 40,
        68, 94, 96
    and activities of experiments, 19–21, 23,
        33–35, 46–50
    bottom-up initiatives and, 7–8
    carbon market development and, 125, 128,
        131, 140–44
    CCX and, 125–26
    and challenges of climate change, 10–12
    cities and, 104, 107, 114–15, 118
    measurement of, 46–50
    megamultilateral development and, 14–15,
        184n44
carbon emissions trading. *See* cap and trade
        systems
Carbon Expo, 160
Carbon Finance Capacity Building Programme
        (CFCB), 94, 166, 172

activities of, 19, 53
in carbon market development, 135, 159–60
and technology deployment in cities, 108–9,
        162
CarbonFix, 135, 166, 173
    activities of, 20, 48–49
carbon markets, 22, 39, 70, 74–76, 85, 89, 94, 96
    carbon as commodity in, 124–25
    Copenhagen meetings and, 4–5, 152
    credit market experimentation in, 144–49,
        198n70
    development of, 8, 26, 75, 123–49, 153, 157,
        159–62, 187n52, 196nn7–8, 196n10,
        196n13, 196n15, 196n23, 197n29, 198n70,
        198n99, 198n101
    emissions trading experimentation in, 140–44
    ethics of, 123, 157
    experimental action cluster on, 123–24,
        134–49, 153, 159, 197n29
    governance models and, 44–45, 47–49, 53, 55
    Kyoto Protocol and, 68, 124–30, 133–38
    multiple venues for, 126–29, 196n10
    offsets and, 124–25, 128–37, 139, 143, 145–46,
        148–49, 187n52, 196n23
    and potentials of experimentation, 153,
        159–62
carbon rationing action groups (CRAGs), 7, 37, 59,
        166, 172
    activities of, 20, 55–56
Carbon Sequestration Leadership Forum, 20, 48,
        50, 166, 172
Carr, Alex, 88–89, 177
Cashore, Benjamin, 157
CCAR. *See* California Action Registry
CCBA. *See* Climate, Community and Biodiversity
        Alliance
CCP. *See under* International Council for Local
        Environmental Initiatives
CDM. *See* clean development mechanism
CERS (Consolidated Emissions Reporting
        Schema), 90
CFCB. *See* Carbon Finance Capacity Building
        Programme
C40 Cities Climate Leadership Group, 28, 36–37,
        74, 84, 91–95, 100–101, 105–6, 110, 166,
        172
    activities of, 19, 51, 53, 85–86, 91, 93–94
    Climate Group and, 85–86, 94–95, 100, 113
    Clinton Climate Initiative and, 91, 93–95, 100,
        109, 116–17, 119
    in Copenhagen, 4, 93, 95
    goals of, 93–95
    origin and launch of, 91–93, 95, 100
    and technology deployment in cities, 108–9,
        113, 116–17, 119
Chambers, John, 114
Chicago Climate Exchange (CCX), 37, 55, 95–101,
        123, 166, 173, 179
    activities of, 20, 56, 98–100